SAFETY AND RISK IN SOCIETY

A CLOSER LOOK AT SAFETY AND SECURITY

SAFETY AND RISK IN SOCIETY

Additional books and e-books in this series can be found on Nova's website under the Series tab.

SAFETY AND RISK IN SOCIETY

A CLOSER LOOK AT SAFETY AND SECURITY

JEFF M. HOLDER
EDITOR

Copyright © 2020 by Nova Science Publishers, Inc.

All rights reserved. No part of this book may be reproduced, stored in a retrieval system or transmitted in any form or by any means: electronic, electrostatic, magnetic, tape, mechanical photocopying, recording or otherwise without the written permission of the Publisher.

We have partnered with Copyright Clearance Center to make it easy for you to obtain permissions to reuse content from this publication. Simply navigate to this publication's page on Nova's website and locate the "Get Permission" button below the title description. This button is linked directly to the title's permission page on copyright.com. Alternatively, you can visit copyright.com and search by title, ISBN, or ISSN.

For further questions about using the service on copyright.com, please contact:
Copyright Clearance Center
Phone: +1-(978) 750-8400   Fax: +1-(978) 750-4470   E-mail: info@copyright.com.

NOTICE TO THE READER

The Publisher has taken reasonable care in the preparation of this book, but makes no expressed or implied warranty of any kind and assumes no responsibility for any errors or omissions. No liability is assumed for incidental or consequential damages in connection with or arising out of information contained in this book. The Publisher shall not be liable for any special, consequential, or exemplary damages resulting, in whole or in part, from the readers' use of, or reliance upon, this material. Any parts of this book based on government reports are so indicated and copyright is claimed for those parts to the extent applicable to compilations of such works.

Independent verification should be sought for any data, advice or recommendations contained in this book. In addition, no responsibility is assumed by the Publisher for any injury and/or damage to persons or property arising from any methods, products, instructions, ideas or otherwise contained in this publication.

This publication is designed to provide accurate and authoritative information with regard to the subject matter covered herein. It is sold with the clear understanding that the Publisher is not engaged in rendering legal or any other professional services. If legal or any other expert assistance is required, the services of a competent person should be sought. FROM A DECLARATION OF PARTICIPANTS JOINTLY ADOPTED BY A COMMITTEE OF THE AMERICAN BAR ASSOCIATION AND A COMMITTEE OF PUBLISHERS.

Additional color graphics may be available in the e-book version of this book.

Library of Congress Cataloging-in-Publication Data

ISBN: 978-1-53618-176-0

Published by Nova Science Publishers, Inc. † New York

Contents

Preface		vii
Chapter 1	CC-Case: Safety & Security Engineering Methodology for AI/IoT *Tomoko Kaneko and Nobukazu Yoshioka*	1
Chapter 2	Safety and Security Challenges Preventing Parents from Enrolling Children in Early Years School in Difficult Circumstances *Nyakwara Begi and Yattani D. Buna*	81
Chapter 3	Safety, Security, and Social Engineering — Thoughts, Challenges and a Concept of Quantitative Risk Assessment *Joachim Draeger*	113
Index		161

PREFACE

This compilation details the challenges in the development of computer systems more complexed by AI/IoT using examples based on systems thinking, as well as introduces how these technologies can aid in safety and security.

The authors carry out a study to establish the safety and security challenges preventing parents from enrolling their young children in schools, with the intention of recommending strategies to address these challenges. This study is guided by the ecological systems theory which explains how factors within the environment influence children's development and education.

The concluding section focuses on the system dynamics paradigm which seems to be a suitable modeling method to describe social engineering situations. Opportunities for future research are also discussed.

Chapter 1 - In this chapter, we will show a new development methodology that can assure the needs of complex systems including IoT and AI by using safety and security technologies in an integrated manner. Specifically, we are considering the integrated use of the following technologies suitable for analysis, implementation, and evaluation of complex systems:

- New system thinking safety technology such as STAMP/STPA
- FRAM, a resilience engineering method

- Expand security functions defined in IT Security Standard Common Criteria (CC)
- SARM that makes it possible to identify comprehensive requirements
- Advanced patented technology such as Scenario Function
- Verification and validation by guarantee cases such as GSN
- System thinking accident analysis method CAST

We aim to develop a safe and secure development methodology for complex systems in the AI/IoT era.

Chapter 2 - Early Years Education (EYE) is a basic right for every child as enshrined in the United Nations Convention on the rights of children. The African Charter on Rights and Welfare of Children, Kenya's Education Act 2013 and Children's Act 2012 emphasize that children's environment should be free from physical and socio-emotional stress that hinders them from accessing EYE. However, research has shown that there are many children in difficult circumstances not accessing EYE owing to safety and security concerns in the environment. In view of the importance of EYE, there was need to carry out a study to establish the safety and security challenges preventing parents from enrolling their young children in schools with the intention of recommending strategies to address the challenges. This study was guided by the ecological systems theory which explains how factors within the environment influence children's development and education. Descriptive survey research design was used; while data was collected using interviews and focus group discussions. The study was done in Marsabit County which is one of the counties in Arid and Semi-arid Lands (ASAL) in Kenya. The target population was school going children and their parents in households in the county. Purposive and random sampling techniques were used to select parents, teachers, area education officers and chiefs. The research instruments used for collecting data were interview schedules and focus group discussions which were analyzed using qualitative and quantitative techniques. Results from data analysis show that several safety and security challenges were preventing parents from enrolling their children in schools. The challenges included:

Long distance from home to school; harsh weather; rugged terrain; salty water; and ethnic conflicts. Some of the recommended strategies to address the challenges include establishment of mobile schools near villages; flexible school routine; avoiding rugged terrain when selecting sites for establishment of villages; boiling and filtering water for drinking; and Organising frequent peace and reconciliation meetings.

Chapter 3 - Social engineering attacks on human operators supervising technical systems can pose a significant risk. Its quantitative assessment faces several challenges, however, which make the application of common assessment methods inappropriate. First, aspects of safety, security, and cognitive psychology have to be included concurrently. Second, the supervised technical system may possess a distinct inherent dynamics. Third, the available knowledge about the cognitive characteristics of the human operator will be limited. The complexity of the situation illustrated in the first point favors model-based considerations, where by the limited knowledge indicated in the third point suggests the usage of comparatively simple models. As it turns out, the system dynamics paradigm seems to be a suitable modeling method to describe social engineering situations. This also takes into account the second point of the list, which leads to a preference for simulations to determine the outcome of specific scenarios.

The representation of possible failures in the model together with their probabilities and criticality values is a prerequisite for a simulation-based calculation of the risk. Interpreting risk as mean expected criticality or, more generally, as probability distribution of outcome criticality leads immediately to a random sampling approach for covering the space of scenarios. The sampling is realized via Monte Carlo simulation runs. Missing knowledge can be integrated by variations of uncertain parameters according to a given probability distribution. Using the central limit theorem, the sampling error can be estimated and confidence intervals given. For assuring the quality of the risk assessment, validation methods for the underlying model are discussed. Using the human driver of a semiautonomous car as an exemplary application demonstrates the practical relevance of the considerations. The importance of understanding the risk assessment is substantiated as prerequisite for its credibility. The chapter closes with a short outlook, in which the proposed concept and opportunities for future research are discussed.

In: A Closer Look at Safety and Security
Editor: Jeff M. Holder

ISBN: 978-1-53618-176-0
© 2020 Nova Science Publishers, Inc.

Chapter 1

CC-CASE: SAFETY & SECURITY ENGINEERING METHODOLOGY FOR AI/IoT

Tomoko Kaneko[*] *and Nobukazu Yoshioka*
QAML Project, National Institute of Informatics, Tokyo, Japan

ABSTRACT

In this chapter, we will show a new development methodology that can assure the needs of complex systems including IoT and AI by using safety and security technologies in an integrated manner. Specifically, we are considering the integrated use of the following technologies suitable for analysis, implementation, and evaluation of complex systems:

- New system thinking safety technology such as STAMP/STPA
- FRAM, a resilience engineering method
- Expand security functions defined in IT Security Standard Common Criteria (CC)
- SARM that makes it possible to identify comprehensive requirements
- Advanced patented technology such as Scenario Function
- Verification and validation by guarantee cases such as GSN
- System thinking accident analysis method CAST

---

[*] Corresponding Author's Email: t-kaneko@nii.ac.jp.

We aim to develop a safe and secure development methodology for complex systems in the AI/IoT era.

Keywords: common criteria (CC), assurance case, IoT, AI, system theory, resilience engineering, STAMP, STPA, CAST, FRAM, GSN, scenario function

1. INTRODUCTION

As the complexity of computer systems increases, the realization of safety and security is important. Safety means being protected from unintended hazards such as accidental mistakes and failures. Security means protecting against malicious threats. Achieving safety and security for those systems requires addressing many challenges.

First, safety is hardware-centric, whereas security is software-centric. Therefore, the respective techniques and countermeasures are different. Safety focuses on the implementation of individual components according to specific standards. Some of the common safety techniques include FTA [1], FMEA [2], and HAZOP [3]. However, those techniques are not totally applicable for System of systems (SoS) or the Internet of Things (IoT) because they cannot deal with interactions between connected components. In addition, as safety-critical industries such as automotive and medical care are increasingly connected to the internet, they also need to consider security. Indeed, security attacks can have a significant impact on lives and health.

On the other hand, security (here we mean IT security) focuses on proactive testing to detect and respond to known vulnerabilities in cyberspace without considering equipment and other devices. Attacks continue to be sophisticated, and the damage caused by information leaks has become even bigger. "Security-by-Design" aims to reduce unnecessary costs by creating this from the planning and concept stage.

We seek to address these challenges by using safety and security techniques in an integrated way.

Methods such as Systems-Theoretic Accident Model and Processes (STAMP), System Theoretic Process Analysis (STPA), Functional Resonance Analysis Method (FRAM), IT Security Standard Common Criteria (CC), and Causal Analysis based on STAMP (CAST) have been researched and implemented with a focus on safety, but the proposed method is applied to both cyber security and software engineering. In addition, the existing method is hardware-centric, but the proposed method conducts research and implementation that extends the scope of application to the requirements of software, systems, society, and specifications. Furthermore, the proposed method aims to not only clarify requirements but also make a methodology that realizes assurance by using each method appropriately. Specifically, new system-thinking safety technologies, such as STAMP/STPA [4] [5] [6] [7], analyze both hazards and threats. In addition, FRAM [8] [9] which is a method of resilience engineering [10] [11] [12], finds safety and security features to resolve identified risks. In addition, it extends the security features defined in the IT Security Standard of Common Criteria (CC) [13] to generate a reusable Protection Profile (PP) suitable for device systems in each domain. This allows one to visualize safety and security features. Furthermore, in the program manufacturing process, the software connection is clarified through state-of-the-art patented technologies such as Scenario Function [14], [15], [16] enabling bug-free software and incapacitating viruses. Each of these steps is recorded by an assurance case and can be used for validation. In addition, CAST (Causal Analysis using System Theory) [6] [7], which is a system-thinking method, can clarify the result of accidents.

In this chapter, we will detail the challenges in the development of complex computer systems in the current phase of AI/IoT. We will also explain each of the aforementioned technologies, and how to achieve integration in order to realize safety & security.

In this chapter, related work is given in Section 2. Section 3 explains the needs for building a safe and secure system in the AI/IoT era, and Section 4 explains what CC-Case [17], [18], [19], the safe and secure system development methodology in the AI/IoT era. Section 5 introduces CC-Case technical elements and examples, Section 6 shows the integrated use of CC-Case technologies and CC-Case for AI/IoT. Section 7, conclusion shows future initiatives and roadmaps.

2. RELATED WORK

2.1. Requirements Analysis Technique

For the extraction of the demand in the software development, there is proposed, a method of using the collected data, the questionnaire interview, the goal analysis, the youth case, the brainstorming i.e., The goal analysis is a technique to analyze and define the target that the demand of the system for development should achieve. The design's adequacy can be confirmed by clarifying the target of the demand and making an agreement with customers, and the validity of the function demand can be confirmed by checking the goal. The extracted demand is classified in the requirements analysis and the competition between demands is solved. In the classification of the demand, to understand the demand, the conceptual model is made. Moreover, the demand is associated with the component of the system in consideration of architecture. The priority level between demands is clarified in the contention resolution, based on mutual agreement with parties concerned. The conceptual model can be classified into a data flow model, a control flow model, an event sequence model, an object-oriented model, the goal aim model, the concept data model, a formal model, the business model, and the business process model. There are i* [20] [21], KAOS [22], and ARM [23], as an example of the goal aim modeling. KAOS is a requirements analysis technique for systematically analyzing the system goal, and the logical analysis based on the formal approach is possible. Several views are offered, and the process from the demand to the design is actualized gradually. On the other hand, i* aims to analyze the demand for the system through the interests between stakeholders. When you compare the range of KAOS with that of i*, i* is examined as the intentional goal between stake-holders with As-Is (current state) and To-Be (the way it should be), KAOS is examined around the system goal. i* is a requirements analysis technique for the analysis of the actor relation from the upper process, and providing the agent oriented analysis. The interest is expected of i* as an effective technique for modeling the stakeholder and the system as an actor, modeling as the dependence between actors. i* develops as a requirements analysis process part of developmental methodology Tropos [24] [25]. However, the effectiveness and limitations of i* are unclear.

To solve the problem, an actor relationship matrix analysis method (ARM) is proposed as a precursor to i* modeling.

For actors A1, .., An, the 2 dimension matrix ARM (A1,... ,An) is defined as follows (Table 1).

The 'A' column and rows correspond with actors, A1, .., An. The element for the i-th row and j-th column represents the intention Ik of actor Ai to actor Aj. The intention includes softgoals, goals, resources, and tasks.

The element for the i-th row and i-th column represents the goals of actor Ai.

ARM(A1,..., An)[i,i] = {Gk|Gk is a goal of Ai, $0 < k < l_i$}

ARM(A1,..., An)[i,j] = {Ik|Ik is an intention of Ai to Aj, $0 < k < l(i,j)$, $i \neq j$}

ARM can classify goals, soft goals, resources, and tasks. ARM can transform to a SD model of an i*framework.

2.2. Traditional Methods of Safety Analysis

Traditional methods of safety analysis have mostly been developed 40 to 65 years ago. Since then, the systems have evolved and gone through a complete transformation. The Fault Tree analysis (FTA) [1], FMEA (Failure Mode and Effect analysis) [2] are traditional hazard analysis methods that analyze hazard factors by using fault trees and impact analysis tables (Figure 1).

Table 1. Actor Relationship Matrix (ARM)

	A1	A2	A3
A1	Goals of actor A1	Intention of A1 to A2	Intention of A1 to A3
A2	Intention of A2 to A1	Goals of actor A2	Intention of A2 to A3
A3	Intention of A3 to A1	Intention of A3 to A2	Goals of actor A3

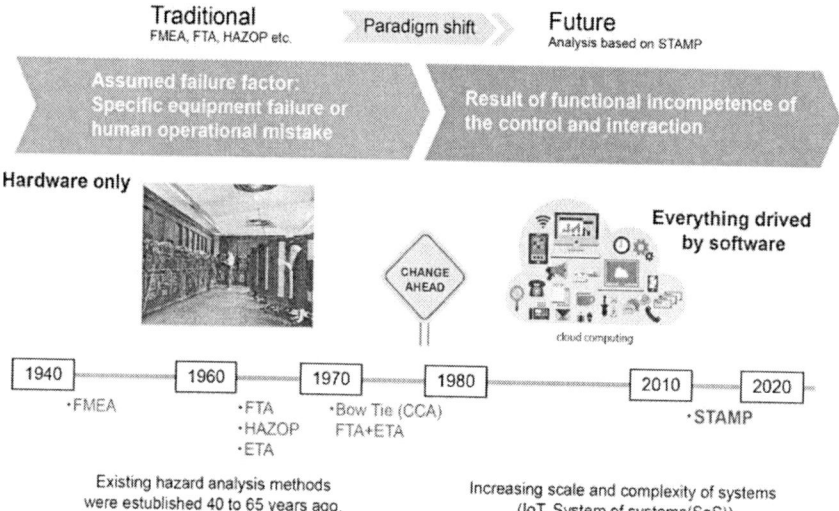

Figure 1. Background of STAMP appearance [26].

It can be applied at the architecture design stage where the system components and failure modes are determined. It is difficult to analyze a single failure of a device or an organization as a hazard factor, although it can be analyzed in a systematic manner by logically forming a branch condition which requires an overall field of view, such as an accident generated from the interaction between components.

2.3. Traditional Methods of Security Analysis

Security analysis techniques include attack trees [27], misuse cases [28], Microsoft Security Development Life Cycle [29], and threat modeling including STRIDE analysis [30]. The security development life cycle [29] provides a detailed data flow diagram and a threat analysis with STRIDE. We extract security requirements in the design stage with an emphasis on ensuring safety by design. A threat tree classification based on STRIDE [30] is also shown. Some HAZOP-based security analysis methods have also been proposed [31].

However, compared with traditional standardized safety analysis methods such as FTA and FMEA, no security analysis method has been standardized. No security design method is widespread in the development field.

As security risks are caused by threats and vulnerability. Security analysis is divided into vulnerability and threat analysis. Threats are caused by attackers, and various assets are exploited and threatened by attackers.

The security requirement is a description of the measures as the threat to the asset. The attack on security is a means to achieve the threat intentionally, and it is necessary to consider the attacker's existence in the security requirements analysis. The purpose of a general requirements analysis is to achieve the stakeholder's intention to the system. On the other hand, the purpose of a security requirements analysis is to obstruct the attacker's intention.

Many security requirements analyses are youth cases, and the goal aim modeling is based on a general requirements analysis technology. There is a technique of misuse case as an example of the youth case to extract the attack to the system and the exception that should be assumed. The misuse case is suitable for expressing the character of the youth case to the function, and extracting non-function demand of security, etc. There is NFR framework as an example of the goal aim modeling, besides the above-mentioned i* and KAOS. The layered structure of general non-function demand (NFR) is defined as NFR type, and the technique used again is by analyzing concrete system requirements. Tondel assumes that the security target, the property extraction, and the threat analysis are common as the most basic element to the security demand technique [32]. There is the Liu method of i* as a security analysis technique for providing with the security target, the property extraction, and the threat analysis.

The Liu method is of i* is one of the security requirements analysis techniques based on the goal analysis. This method adds the following security requirements analysis processes to the i* framework that is the requirements analysis technique of a general function:

1. Attacker Identification: The attacker as the threat on security is specific as the actor.
2. Malicious Intent Identification: The attacker's intention (malice) is specific as the goal.
3. Vulnerability Analysis: The property with vulnerability is specific.
4. Attacking Measure Identification: The method of attack is specific as attacker's task.
5. Countermeasure Identification: It is specific as the task of the actor becomes the target for measures against the method of attack. Then the security requirements analysis based on the goal aim is achieved [33] [34]. However, the problem of i* described in section 2.1 exists even though the Liu method of i* enhances i*. Kaneko has proposed SARM as a method to solve these problems [35] [36]. SARM is explained in detail in Section 5.1.

2.4. STAMP and Its Related Methods

Modern embedded systems are becoming gradually larger and more complex due to the interaction among connected elements in addition to the advanced functionality of each element.

To ensure the safety of those complex systems, Leveson proposed Systems-Theoretic Accident Model and Processes (STAMP) and its safety analysis application, System Theoretic Process Analysis (STPA). STAMP [4] is not an analysis method. Instead it is a model or set of assumptions about how accidents occur.

STAMP is an alternative to the chain-of-failure-events (or dominos or Swiss cheese slices, all of which are essentially equivalent).

1. Domino Model

The series of cause and effect (the following causes) is called the Domino model.

If you hold your hand somewhere in this domino effect, you can avoid the accident.

Each technique of the accident analysis says the root cause analysis stands in this idea.

2. Swiss Cheese Slices Model

The defense wall and the leak is like the holes in the cheese. It is called the Swiss cheese slices model

It becomes an accident when the holes overlap and it foreseen. This is dealt with by blocking the individual holes.

It underlies the traditional safety analysis techniques (such as Fault Tree Analysis [1], FMEA [2], and HAZOP [3]). Just as the traditional analysis methods are constructed on the assumptions about why accidents occur in a chain-of-failure-events model, new analysis methods can be constructed using STAMP as a basis. Note that because the chain-of-failure events model is a subset of STAMP, tools built on STAMP can include as a subset, all the results derived using the older safety analysis techniques.

STAMP is an accident model based on system theory, and STPA [5] is a typical method based on the STAMP model, and the hazard analysis is performed.

Many of the system accidents are not caused by the failure of the components, but the interaction of the control elements (control element and the controlled element) for safety in the system. The mechanism is explained by focusing on the element (component) and the interaction (Control action). The cause of "action not working" is limited by having the viewpoint of being equal to "inappropriate action of the control action."

Some advantages of using STAMP are:

- It works on very complex systems because it works top-down rather than bottom up.

- It includes software, humans, organizations, safety culture, etc. as causal factors in accidents and other types of losses without having to treat them differently or separately.
- It allows creating more powerful tools, such as STPA, accident analysis (CAST) [6], identification and management of leading indicators of increasing risk, organizational risk analysis, etc.

Because STAMP applies to any emergent property, STPA can be used for any system property, including cybersecurity.

As a process, STAMP uses specifications, safety guide design, design principles, system engineering, risk management, management principles, and regulation of organizational designs (Figure 2). Based on the STAMP model, an accident/event analysis (CAST: causal analysis based on STAMP), hazard analysis (STPA), early concept analysis (Steca: systems-theoretic early Concept analysis), systematic/cultural risk analysis, leading indicator identification, security analysis (STPA-Sec) is presented. An accident/event analysis (CAST) is a method of analyzing an event since an accident occurred, and STPA-Sec is the security version (Figure 2). STPA-SafeSec has been proposed as a method of integrating safety and security [37] [38].

2.5. Resilience Engineering and FRAM

Safety engineering is expanded by STAMP, assuming that the cause of an accident is not only caused by a component failure but also caused by a failure of the mechanism that normally prevents the accident.

Hollnagel [10] [11] [12], on the other hand, extended further, assuming that accidents could arise not only from failure but also from success. In systems where it is difficult to assume a fixed mechanism to ensure safety such as artificial intelligence that changes behavior non deterministically according to the environment, it is necessary to narrow down in advance what functions control safety. In order to model such a system for safety analysis, it is necessary to find a mechanism that provides safety in the first place.

"The traditional view of safety, called Safety-I, has consequently been defined by the absence of accidents and incidents, or as the 'freedom from unacceptable risk.' As a result, the focus of safety research and safety management has usually been on unsafe system operations rather than on safe operation. In contrast to the traditional view, resilience engineering maintains that 'things go wrong' and 'things go right' for the same basic reasons. This corresponds to a view of safety, called Safety-II, which defines safety as the ability to succeed under varying conditions" [11].

Resilience is the intrinsic ability to succeed under changing conditions. Resilience engineering is the design methodology of resilience. The objective of risk management is not reduction of risks, but enhancement of the ability to suppress system performance variability under changes, disturbances, and uncertainties.

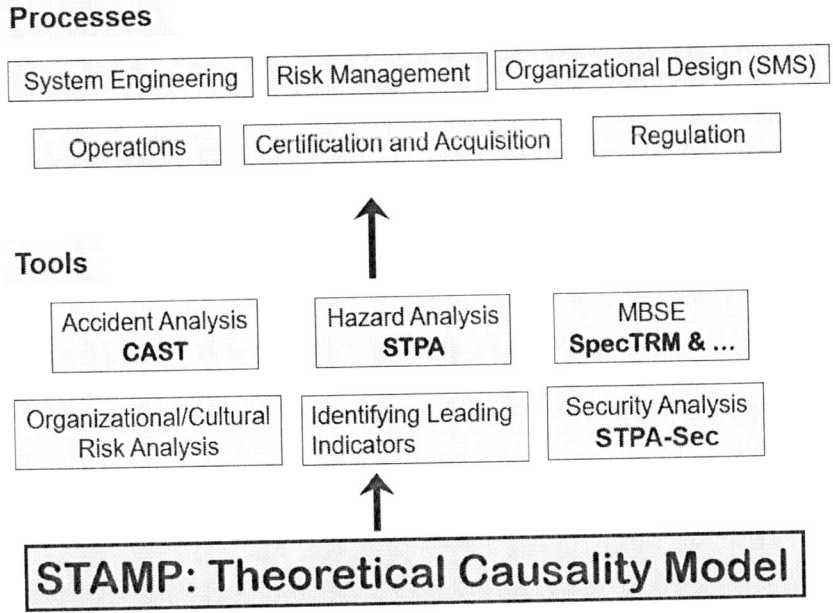

Figure 2. STAMP related methods [7].

Resilience engineering emphasizes these four abilities, (1) Anticipation – get ready for long-term threats and changes, (2) Monitoring – watch system states and find out clues of threats, (3) Responding – take immediate actions to regulate function variability, (4) Learning – learn from good as well as bad consequences.

A modeling method developed based on this concept is the Functional Resonance Analysis Method (FRAM) [8] [9]. This is a way to describe outcomes using the idea of resonance arising from the variability of everyday performance. To arrive at a description of functional variability and resonance, and to lead to recommendations for damping unwanted variability, a FRAM analysis consists of four steps:

Identify and describe essential system functions, and characterize each function using the six basic characteristics (aspects). In the first version, only describe the aspects that are necessary or relevant. The description can always be modified later.

Characterize the potential variability of the functions in the FRAM model, as well as the possible actual variability of the functions in one or more instances of the model.

Define the functional resonance based on dependencies/couplings among functions and the potential for functional variability.

Identify ways to monitor the development of resonance either to dampen variability that may lead to unwanted outcomes or to amplify variability that may lead to wanted outcomes.

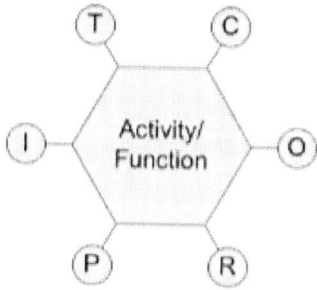

Figure 3. The six basic characteristics of FRAM [9].

The six basic characteristics (aspects) are Input (I: Input that triggers the start of the function), Precondition (P: Input that is a precondition of starting the function), Resource (R: Input that is the resource required to the implementation of the function), Time (T: Time information that restricts the implementation of the function), Control (C: Control inputs that change the way functions are performed) Output (O: Function output) (Figure 3).

2.6. Assurance Case

Assurance case, which is defined in ISO/IEC15026 part2, is a method for describing a system's critical security level. Standards are proposed by ISO/IEC15026 [39] and OMG's Argument Metamodel (ARM) and [40] Software Assurance Evidence Metamodel (SAEM) [41]. ISO/IEC 15026 specifies scopes, adaptability, application, assurance case's structure and contents, and deliverables. Minimum requirements for assurance case's structure and contents are: to describe claims of IT products and systems properties, systematic argumentations of the claims, evidence and explicit assumptions of the argumentations; to structurally associate evidence and assumptions with the highest-level claims by introducing supplementary claims in the middle of a discussion (Figure 4). One common notation is Goal Structuring Notation (GSN) [42], which was widely used in Europe for over ten years to verify system risk and validity after identifying risk requirements. Contents of GSN is shown below (Table 2).

The minimum requirements for assurance case's structure and contents are: to describe claims of system and product properties, systematic argumentations of the claims, evidence and explicit assumptions of the argumentations; to structurally associate evidence and assumptions with the highest-level claims by introducing supplementary claims in the middle of a discussion.

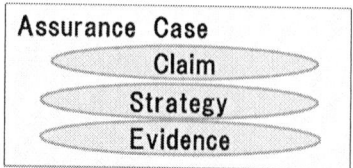

Figure 4. Minimum requirements for assurance case.

Table 2. The representation of GSN

Contents	Figure	Explanation
Goal (Claim)	▭	To describe claims of system and product properties
Stragety (Argumentation)	▱	Systematic argumentations of the claims
Context (assumption)	▢	Explicit assumptions of the argumentations
Undefind	◇	Undefind claims and explanation
Evidence	○	Evidence of the argumentations

Lipson and others proposed a method to create a Security Assurance case [43]. They described that the common criteria provides catalogs of standard Security Functional Requirements and Security Assurance Requirements. They decomposed security assurance cases by focusing on the process, such as requirements, design, coding, and operation. The approach did not use the Security Target structure of the CC to describe the Security Assurance case.

Alexander, Hawkins and Kelly reviewed the state of the art on the Security Assurance cases [44]. They showed the practical aspects and benefits to describe the Security Assurance case in relation to security target documents. However they did not provide any patterns to describe the Security Assurance case using CC. Kaneko, Yamamoto and Tanaka proposed a security countermeasure decision method using assurance case and CC [45-48]. Their method is based on a goal oriented security requirements analysis. Although the method showed a way to describe security assurance case, it did not provide any Security Assurance case graphical notations and the seamless relationship between security structure and security functional requirements.

2.7. Common Criteria

2.7.1. What is Common Criteria (CC: ISO/IEC15408)?

Common Criteria (CC: equivalent to ISO/IEC15408) [48] specifies a framework for evaluating reliability of the security assurance level defined by a system developer.

As an international standard, CC is used to evaluate reliability of security requirements of functions built using IT components (including security functions). CC establishes a precise model of Target of Evaluation (TOE) and the operation environment. Based on the security concept and relationship of assets, threats, and objectives, CC defines ST (Security Target) as a framework for evaluating TOE's Security Functional Requirement (SFR) and Security Assurance Requirement (SAR) (Figure 5).

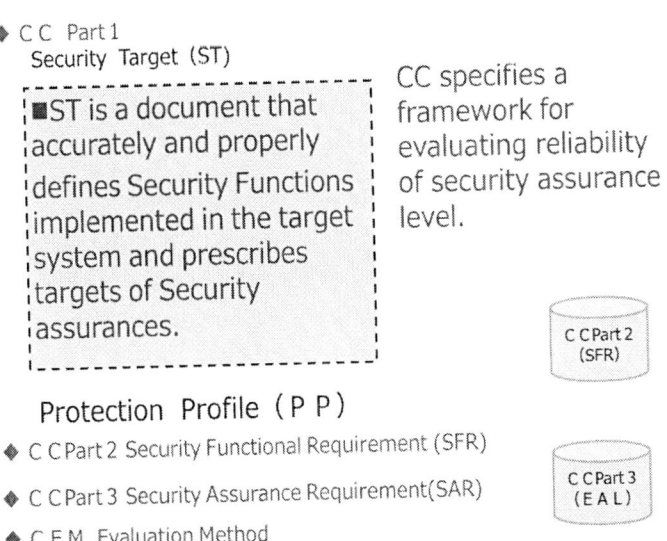

Figure 5. CC configuration.

ST is a document that accurately and properly defines security functions implemented in the target system and prescribes targets of security assurance. ST is required for security evaluation and shows levels of adequacy in TOE's security functions and security assurance.

*Based on CC Recognition Arrangement (CCRA), which recognizes certifications granted by other countries' evaluation and authorization schemes, CC accredited IT products are recognized and distributed internationally.

2.8. Scenario Function

In the IoT era, all devices are controlled by software. To realize safety and security, secure new software is essential.

The change in the way the program itself is created must solve today's critical software challenges. Secure software needs to be bugless, also it is desirable not to be able to be defeated by attacks by computer viruses. In addition, there is a need for something that can avoid the rising costs of security measures.

We introduce patents by Negoro, who has been researching for many years to solve program problems. This patent has acquired Japan No. 5992079, No. 60869977 and US Pat. No. 10,235,522[14], [15], [16].

The present invention solves many problems of the past, and provides a fundamental solution to a computer virus problem, that is, a program definition structure for autonomously solving a virus problem and a program having the same structure. It is an object of the present invention to provide a storage medium having the same structure and a method for autonomously solving a virus problem.

Table 3 shows the definitions of a synchronous algorithm, a logically coupled program, an asynchronous algorithm, and a data coupled program.

Table 3. Difference between Current software program and Scenario Function

	Current software program	Program proposed by the inventor
Static structure	Logically coupled program: LCP	Data binding program: SF
Runtime algorithm	Non Synchronization algorithm: NSA	Synchronization algorithm: SA
Characteristic	Programs that cannot be verified	A new static structure designed to create a complete dynamic algorithm at runtime

The difference between the two is, for example, "asynchronous processing is such that words that are uttered once in a person's conversation are not particularly post-processed and are terminated when they are spoken. Such an aspect is defined as an infinite phenomenon problem. The current software program is called an asynchronous algorithm because it is processed asynchronously for each module (partial program whose execution validity is unknown) as in the case of words. Asynchronous algorithms are logically linked, so they are called logically coupled programs. On the other hand, the synchronous algorithm transcends the module unit, grasps the meaning of the data of the program to be processed, and allows the same loop to be repeated until the processing of the entire program (synchronization processing) is completed. It is SF that establishes this synchronous algorithm.

The technology will be briefly described as follows:

1. The source structure of a logically coupled program (hereinafter LCP) causes an asynchronous algorithm (hereinafter NSA) to occur at runtime.
   What is LCP verification?
   This is a success/failure judgment of the legitimacy of NSA that cannot be implemented originally.
2. The correctness of the program should be judged by the success or failure of the entire sufficient condition of the program captured by the algorithm generated by the program at the time of execution.

3. Although it is possible to define LCP partialization, the entire sufficient condition of NSA (i) generated by the program (hereinafter LCP (i)) at the time of execution cannot be grasped only by NSA (i). NSA (i) ∪ NSA can only be captured. Although this is possible to define LCP (i),

   This is the reason why the success or failure of the algorithm during execution is unknown.

4. The structure of the source of the LCP has not been a structure that can determine the legitimacy of the NSA since its birth.

   Therefore, the LCP operates while the legitimacy of the NSA remains unknown and has an illegal value that develops during execution.

5. As a result, NSA causes LCP, and NSA (i) causes LCP (i) to cause unsolvable program problems.

6. The SF is a program designed by the SF inventor after over 40 years of research so that the source structure generates a synchronization algorithm (hereinafter referred to as SA) at the time of execution.

7. SF is a program problem, for example, an illegal value caused by syntax at runtime,

   Capture only the information necessary to establish the original algorithm that satisfies the intent to be obtained while autonomously capturing the illegal values sent by the virus with the SA mechanism and eliminating them autonomously, and autonomously correlating them is possible.

8. As a result, the source of the SF can be obtained by defining it according to the universal definition rule by the SF as a structure for establishing the SA at the time of execution.

LCP has program issues since its birth, such as bug issues, virus issues that have become popular in recent years, etc.

There are problems that cannot be completely enumerated. Today's officials have overlooked these with superb procedural techniques, but in the SF world, these problems can be ignored by the SF system.

3. Needs of Safety and Security for AI/IoT Era

This section describes the needs for safe and secure system construction in the AI/IoT era. Specifically, Section 3.1 describes the definition of safety and security, the difference between the two, and the issues that it faces. Section 3.2 describes safety and security needs, especially from the perspective of IoT. Section 3.3 describes AI safety and security issues.

3.1. Safety and Security Challenges

You can define safety and security by the difference between what kind of opponent you want to protect from and what you want to protect. This chapter defines in terms of what kind of opponent you want to protect from, and safety shows unmalicious hazards such as accidental mistakes and failures, while security indicates malicious threats. (Safety factors are called hazards and security factors are called threats.)

Business equipment and systems can potentially cause harm to a user's body or property due to malfunctions or third-party attacks, such as software defects or vulnerabilities. If harm occurs, the business impact will be great, such as general damages and broken equipment. Safety and security risks can be assessed based on the ease of hazards and threats and the severity of the damage. If the accuracy rate of damage is serious but close to zero, the risk is reduced, and the risk increases if minor damage spreads over the network. Safety enhancements ("safety features") are often controlled by software, so if a security threat affects the software of other devices through the network, a wide range of safety features is necessary. The risk is not measurable.

IoT requires proactive action because hazards and threats can spread extensively and pose a significant risk to the business of the enterprise. Recently, everything is connecting through the Internet, so security threats lurk everywhere.

Therefore, by creating connected devices and systems technology to assure safety and security is needed.

3.1.1. What is the Difference between Safety and Security?

We can show differences between safety and security in several ways (Table 4). Safety protects human life, property (house, etc.), whereas security protects information confidentiality, integrity, availability, etc. Safety is affected by misuse or equipment malfunction, whereas security is affected by an intended attack.

Safety issues are easy to detect because they often appear as an accident, whereas security causes much damage from eavesdropping and intrusion. Regarding frequency of occurrence, safety is stochastic because it can be treated as a probability of occurrence, whereas security is non stochastic because of a person's intended attack. Safety can be helped by risk analyses and countermeasures at design time, whereas security measures need to be created. New attack methods have developed over time, requiring continuous analysis and countermeasures. Safety is hardware or people-centric, whereas security is software-centric. Safety is dealing with whole exhaustive coping, whereas security is an occasional best effort.

There are differences in historical backgrounds. In Japan, a country of manufacturing, many people are involved in the safety of automobiles, home appliances, medical devices, etc.

Safety is important because it affects human life and health, and has a long history. Safety starts with people's mis-response, and has expanded to the cooperation of the person and the machine.

In recent years, by connecting via the Internet, automobiles and medical devices have an increased security threat, having been attacked by remote control, and manufacturers are beginning to respond.

There is a change in the purpose of attacks through the Internet. From early pranks, unauthorized data access has been replaced by attacks to get money. The black market of cyber threats is said to have become huge and reached fifteen billion dollars. AI attacks will surely become even bigger.

Table 4. Differences of Safety & Security

Points of differences	Safety	Security
Protection target	Human life, property (house, etc.)	Information confidentiality, integrity, availability, etc.
Cause	Reasonably foreseeable misuse, equipment malfunction	Intended Attack
Damage detection	Easy to detect because it appears as an accident	Many damage, such as eavesdropping and intrusion, is difficult to detect
Frequency of occurrence	Stochastic because it can be treated as a probability of occurrence	Non-Stochastic because of a person's intended attack
Timing of countermeasures	Available to respond by risk analysis and countermeasures at design time	New attack methods are need developed over time, requiring continuous analysis and countermeasures
Focus point	Hardware or people-centric	Software-centric
Workaround	Whole exhaustive coping	Occasional best effort
Historical perspective	Safety has long history and many standard, traditional analysis method	With the advent of computers and the Internet, it became necessary
Standard	There are many standards for each domain and it is mandatory to follow the standards.	Management standards are generally audited, but IT security international standards for development and design are not sufficiently widespread.
Risk analysis	There are many traditional methods such as FTA, FMEA, and HAZOP.	Safety methods or software engineering with security added have been devised, but not universal.

3.1.2. Safety and Security by Design

Security measures have currently focused on dealing with vulnerabilities from the operational stage.

However, as security threats are becoming more likely to cause significant damage in the Age of IoT, security needs to be built in advance from an earlier stage. The cyber Security Center of the Cabinet Office in Japan (NISC) defines "Security by Design" as "measures to ensure information security from the planning and Design Stage" [49].

In the "General Framework for Secure IoT Systems" [50], it is an important concept that is raised as a basic principle as described below. "Security by Design" should be a fundamental principle in designing, deploying and operating IoT systems. A framework is needed to confirm and verify the fundamental principles prior to deployment." It is required to create security from the early stages of planning and requirements definition processes and design processes (Figure 6).

There is a term "security by design," but safety by design is natural. In other words, the main response procedure is different. Safety is typically built from the early stages of planning and requires attention in the design process. Next, safety is audited and to meet standard requirements.

Security focuses on incident response. This is because new threats are constantly generated that were not anticipated during product planning. Information security management systems are often audited, but Common Criteria, an IT security evaluation standard for system products, is used only for specific products. For these reasons, creating products that meet audit requirements from the planning stage is not very common.

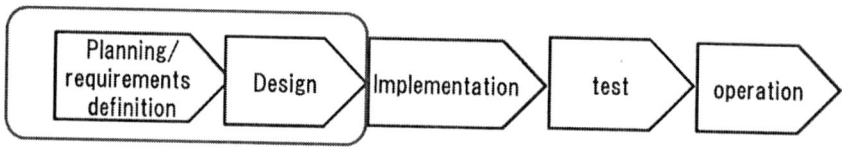

Figure 6. Concept of Security by Design.

For the real integration of safety and security requirements in the design process, both safety and security requirements should be considered at the same time (Figure 7). By incorporating security design requirements into the existing safety design process, the development of highly reliable systems could be achieved more effectively [51]. The current status of international standards (ISO/IEC) and de facto standards should be taken into account.

Furthermore, the safety of IoT is described as follows. "Internet of Things (IoT) systems consist of connected things and networks and thus should be regarded as an integrated system of IT with physical components. It is important to ensure physical safety in addition to existing information security measures. It is essential that IoT systems are designed, developed and operated under the principle of "Security by Design," while looking ahead to the future where many individual systems are interconnected with new vulnerabilities possibly introduced [50]." In this way, from the viewpoint of CPS (Cyber Physical System), it is important to ensure safety in addition to ensuring conventional information security, so a development methodology that can use safety and security in an integrated manner is required. We build on "General Framework for Secure IoT Systems" to realize Security by Design in IoT. Since the era of IoT and security threats have been likely to cause a great deal of damage in cyber and physical systems.

Figure 7. Safety and Security Requirements.

3.2. IoT Features and Requirements

What are the characteristics and demands of IoT? In this section, explaining the guidelines of "IoT Development Guidelines" and "IoT High Reliability Functions" will serve your references for IoT needs.

These IoT guidelines show the requirements for the IoT development, however, it does not show how these are realized. Since it only shows the needs of IoT, we want to present the seeds that are the realization method in the next section.

3.2.1. IoT Development Guidelines

"IoT Safety/Security Development Guidelines" [52] identify 17 guidelines which management, developers and operators should consider developing to operate a safe and secure IoT (Table 5).

Table 5. IoT high reliability requirements and functional requirements

Major item		G/L#	Guidelines
Policy	Making corporate efforts for the Safety/Security of the Smart-society	1	Formulating the basic policies for Safety/Security
		2	Reviewing systems and human resources for Safety/Security
		3	Preparing for internal frauds and mistakes
Analysis	Recognizing the risks of the Smart-society	4	Identifying the objects to be protected
		5	Assuming the risks caused by connections
		6	Assuming the risks spread through connections
		7	Recognizing physical security risks
Design	Considering the designs to protect the objects to be protected	8	Designing to enable both individual and total protection
		9	Designing so as not to cause trouble in other connected entities
		10	Ensuring consistency between the designs of safety and security
		11	Designing to ensure Safety/Security even when connected to unspecified entities
		12	Verifying/validating the designs of safety and security
Maintenance	Considering the designs to ensure protection even after market release	13	Implementing the functions to identify and record own status
		14	Implementing the functions to maintain Safety/Security even after the passage of time
Operation	Protecting with relevant parties	15	Identifying IoT risks and providing information after market release
		16	Informing relevant business operators of the procedures to be followed after market release
		17	Making the risks caused by connections known to general users

The guidelines also allow applications across various sectors of the IoT. In addition, the guidelines are the first development guidelines in Japan for safe and secure IoT devices and systems. There are needs for safer and more secure systems based on 17 guidelines that CEOs, management, developers and operators should consider.

We developed the guidelines as measures of risks. The development guidelines consist of 5 categories and 17 guidelines. Basically, there are guidelines that developers should consider. However, CEO and management should consider guidelines of the 1st category. It is very important since it does not continue being right even when CEOs or managers do not understand the importance.

3.2.2. IoT High Reliability Functions

In Japan, there were some requests to embody the technical measures of "development guidelines." The guidance further embodies technical parts of development guidelines. Guidance for Practice Regarding "IoT Safety/Security Development Guidelines" [IoT High Reliability Functions] [53] shows technical parts of development guidelines. It describes the requirements to be considered from the design phase and specific examples of the IoT high reliability functions. Furthermore, it includes inter-sector use cases and examples of risks, threats, and countermeasures applying IoT high reliability functions to those cases. In the guidance, we take the life cycle from the start of using IoT until end of use into consideration. Placement on edge, fog and cloud layers is also considered. This picture means that necessary functions are included during the design phase, regarding measures for maintenance and operation.

Table 6 summarizes the IoT high reliability requirements and functional requirements. We classify the use of the system through five viewpoints; start; prevention; detection; fault tolerance & recovery; and termination.

The classification is as shown in the figure on the previous page (Figure 8). We show the IoT high reliability requirements and functional requirement contents required by each. For functional requirements, numbers indicates necessary functions. For example, there is a risk of information leakage from things discarded, so it is necessary to erase data correctly when you end use.

Classified through five perspectives on maintenance and operation are: Initiation, Prevention, Detection, Fault tolerant & Recovery, Termination.

For the functional requirements shown in Table 7 on the previous page, the guidance introduces 23 functions. Although function names are general, their contents consider IoT. For example, device authentication, lightweight cryptography, whitelist anti-virus measures, etc. Also, the erase function is explained. It is a useful function at termination.

Table 6. IoT high reliability requirements classified through five perspectives

IoT high reliability requirement		Corresponding IoT high reliability function	Corresponding IoT high reliability function
Intiation	[#1] Safely/security/ reliability are attained when the target IoT is introduced or when the use of IoT is started.	[#1] Initial settings are configured properly, and are confirmed as appropriate.	1, 2
		[#2] Can confirm that permission is granted when beginning to use the service.	3, 4
Prevention	[#2] Abnormalities during operation can be prevented.	[#3] Predictive signs of abnormalities can be identified.	5, 6, 7, 8, 9
		[#4] Functions and assets that should be protected can be protected.	4, 5, 6, 10
		[#5] Preparations can be made for abnormalities.	11
Detection	[#3] Early detection of abnormalities during operation is possible.	[#6] Occurrences of abnormalities can be monitored and notified.	12, 13
		[#7] Events can be logged for identification of causes of abnormalities.	5, 6
Fault tolerance & recovery	[#4] Operation can be continued and early recovery is possible even when there is an abnormality.	[#8] Configurations can be identified.	14
		[#9] Operation can be continued even when there is an abnormality.	8, 15, 16, 17
		[#10] Early recovery is possible when there is an abnormality.	11, 18, 19, 20
Termination	[#5] Safety/security/ reliability can be ensured even when the use of system/service is terminated or when the system/service is no longer available.	[#11] Use of system/service can be terminated autonomously or suspended.	18, 21, 22
		[#12] Data can be erased.	23

Table 7. IoT high reliability 23 functions

	IoT high reliability functions				
1	Initial setting function	9	Antivirus function	17	Redundant configuration function
2	Setting information confirmation function	10	Encryption function	18	Suspend function
3	Authentication function	11	Remote update function	19	Recovery function
4	Access control function	12	Monitoring function	20	Fault information management function
5	Log collection function	13	State visualization function	21	Operation protection function
6	Time synchronization function	14	Configuration information management function	22	Lifetime management function
7	Predictive function	15	Isolation function	23	Erase function
8	Diagnostic function	16	Degenerate function		

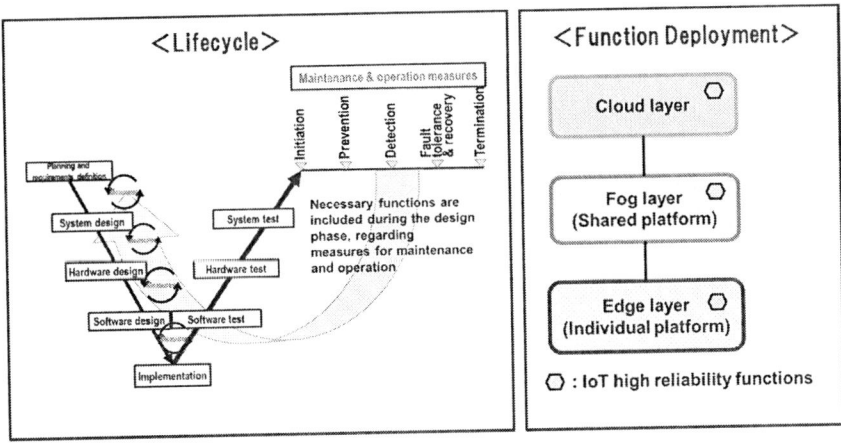

Figure 8. Lifecycle requirements of IoT.

3.3. AI Features and Requirements

What are the characteristics and demands of IoT? In this section, explaining a proposal submitted by a government agency and some reports are helpful in considering AI issues.

Japan Science and Technology Agency's (JST) strategic proposal issued in December 2018 states as follows. "AI software engineering has been developed and established as a new academic field, and as a research and development subject for Japan to secure international competitiveness with this technology, systematization of AI software engineering and four important technical challenges as basic research (machinery, quality assurance of learning itself, ensuring safety as a system, countermeasures for black box problems, engineering framework for solving problems efficiently). Ensuring system safety and reliability cannot be achieved by a single technology, but multifaceted and comprehensive efforts are indispensable. You should work with a good view and balance while keeping the overall picture. An approach to systematize knowledge gained from practice, especially in the industrial world, is expected to progress. However, an approach to academic research related to theory and principle is indispensable to respond to the paradigm shift."

4. WHAT IS CC-CASE?

Based on the description in Chapter 3, in an integrated system life cycle, we present a method for solving problems whenever it is applied to AI and IoT using safety, security and the technology group.

The technology group to be solved is one that can contribute to the solution of problems especially for complex systems among the technologies listed in Section 2. Each technology has its origins in safety, security, or software engineering, but it is not used as it is, but if it is a safety technology, it is diverted to security, and if it is a security technology, it is diverted to safety. If it is software engineering technology, it will be diverted as a technology related to safety and security of the system itself. Research that can apply to AI and IoT and ingenuity to use in an integrated manner is necessary.

The authors named the methodology to accomplish these as CC-Case [17-19]. CC-Case was originally born as a development methodology for security requirements and assurance. The explanation is introduced in a public institutional publication [51] as "When obtaining Common Criteria certification, "CC-Case" is proposed as a method for determining security specifications using Assurance Case. In Common Criteria certification, security design specification documents for target products, Security Target (ST), are required. In CC-Case, ST creation (setting and validation) processes are described while Common Criteria certification standards are modeled and positioned as contexts. This makes creation of ST that complies with Common Criteria certification standards and confirmation of validation by evaluation bodies easier." However, as described in section 3.1.3, in IoT, in addition to ensuring conventional information security, it is important to ensure safety, so we are developing a methodology that can assure both safety requirements and security requirements.

CC-Case is a development methodology that integrates safety and security technologies that can solve the needs raised in Chapter 3. The definition of CC-Case, its purpose, scope and merit, characteristics of the method of application at each process step are described below. In addition, Table 8 shows how to use each technology that constitutes CC-Case, divided into corresponding technologies, base, original characteristics, main scope of application, and implementation stage. The significance of using CC-Case technical elements in an integrated manner is explained.

4.1. Definition and Purpose of CC-CASE

4.1.1. Definition of CC-Case

CC-Case is a system development method that realizes safety and security requirements and assurances through integrated use of various modeling methods, technologies, and standard processes. CC-Case has a two-layer structure consisting of a logical model and a concrete model. The logical model presents the process of creating an assurance case logically, and the concrete model describes the actual case.

CC is a logical layer of the two layers that make up CC-Case. It means the common standard. The authors originally intended to use the IT security standard process based on CC (Common Criteria: ISO/IEC15408), which is a standard process for IT security. However, since the CC-Case framework of "development methodology that can assure safety and security requirements in the AI/IoT era" was established, CC means not only IT security standards but also common standard processes according to each domain. Ultimately, it aims to establish common criteria for each domain.

CASE, a keyword in the automobile industry, indicates "C" for connectivity, "A" for autonomous (autonomous driving), "S" for shared, and "E" for electric. Compared with this, the Case of CC-Case indicates C: Connected, a: assurance, s: system, software, service, society, scenario, and e: evidence.

4.1.2. Scope of CC-Case

The scope of CC-Case includes all stages of the life cycle from the requirement, design, implementation, test, maintenance, and operation stage. By using technical elements suitable for each stage, the life cycle requirements of complex systems are assured.

4.1.3. Purpose of CC-Case

The purpose of CC-Case is to clarify the requirements of advanced and complex systems such as AI and IoT, and to realize systems engineering that can ensure safety and security throughout the life cycle.

4.1.4. Targets and Merits of CC-Case

Targets of CC-Case are IT products or systems, software that makes it up, people who use it, and organization. CC-Case forms an agreement between the customer and the developer, analyzes the requirements including safety and security of the system, incorporates the requirements into functional requirements, visualizes the functions, and leads to assurance. This visualization is also implemented in software programming. To realize these concepts, new modeling methods and technologies will be applied. To make these frameworks that can be applied dynamically, we aim to establish common criteria according to the domain.

By extension from requirement to life-cycle process, CC-Case has merits as mentioned below.

1. It makes it easier to treat additionally for risks to change. Security risks change incessantly because an invisible enemy exists, and an unexpected new menace occurs.
2. It makes it possible to improve the development method, which is issue of assurance case. By defining development processes at each stage, CC-Case would become development method with assurance of life-cycle.
3. CC-Case at requirement stage has only assurance as the future expectation. However, CC-Case of life-cycle extension has assurance with real products.

4.2. Technical Elements of CC-Case

The technical elements and processes to be used are as follows (Table 8).

- Technical elements included in CC-Case cover the system life cycle and are used in cooperation with each other. Because of this cooperative use, higher safety and security can be realized.
- Each technical element is a safety or security method, but it aims to realize both safety and security by expanding it to characteristics different from its origin.

4.3. How to Use CC-Case Technical Elements in an Integrated Manner

To establish CC-Case, do you need a design and a way to integrate and use safety and security?

4.3.1. Why do We Need an Integrated Development Methodology?

As explained in Section 3, in the AI/IoT era, it becomes a cyber-physical and complex system that needs to be considered even for physical systems that cannot be handled by safety, security, and software engineering methods, and requires a new approach.

4.3.2. What Should We do?

It has the needs explained in Section 3.

4.3.3. How Should We do?

We should use technology that achieves both safety and security in an integrated manner. Based on the experience of the first author who has been working as a system engineer and academic researcher for many years, the factors are process, risk analysis, function, and assurance case.

These are preferably the methods listed in Table 8 that originate in safety, security, and software engineering. Each of these techniques is intended to be used dynamically in a suitable process, in conjunction and in response to changes. A specific image is shown in Figure 9.

CC-Case uses industry standards as a process. (This standard uses STAMP/STPA to improve the process to ensure safety and security.)

For risk analysis, STAMP/STPA is used to comprehensively analyze and deal with risks throughout the system.

As for the functions, hidden hazards and threats are identified by using FRAM for the relationship between functions. The requirements extracted in graphic form such as FRAM can be converted to SARM tabular form, increasing its completeness and allowing one-to-one analysis and status monitoring of each function. Tabular status monitoring by SARM is suitable for monitoring a large number of sensors used in automated driving.

The IoT high-reliability function and Protection Profile (PP) are used to formalize the function. When seeking highly reliable software, programming uses scenario functions that achieve completeness and virus neutralization.

Table 8. The technical elements of CC-Case

No	Corresponding Technologies	Base	Original Characteristics	Main Scope of Application	Implementation Stage
0	CC-Case: Safety & Security Engineering Methodology of complexed systems for AI/IoT era, CC-Case includes below technologies (1-8)	Integration of technologies (1-8)	Safety · Security	Integrated application of 1-8: Systems engineering Safety and Security	Entire system life cycle
1	Illustration of hierarchical system by logical layer and physical layer by STAMP S&S	System Theory, STAMP	Safety	Clarification of required specifications (top-down type)	Request (top-down)
2	Safety and security integrated analysis by STPA	System Theory, STAMP	Safety	Risk analysis	Request (top-down)
3	Accident analysis with CAST	System Theory, STAMP	Safety	Accident result analysis	Operation
4	Analyzing the relationship of fluctuating systems using FRAM (Functional Resonance Method)	Resilience engineering	Safety	Clarification of required specifications (bottom-up type)	Request (bottom-up)
5	Quality assurance by GSN	Assurance Case	Safety	Verification and validation	Test
6	Componentization of functional design by PP (Common Criteria)	IT Security Standard	Security	Software function design	Design
7	Clarification of data relationships and virus disabling using Scenario Function	Lyee Theory	Safety, Security	Software programming	Software programming
8	Comprehensive analysis of relationships between actors and functions using SARM	Goal-oriented model	Security	Requirement/design details and countermeasures	Requirements (top-down)/design

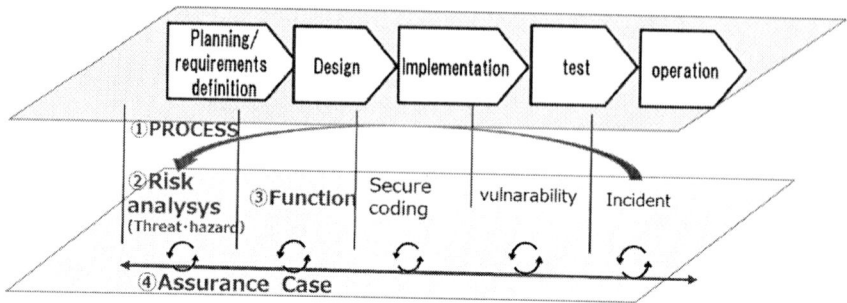

Figure 9. Specific image of CC-Case Integration.

As an assurance case, GSN is used dynamically for each component. This GSN will be used in two layers as shown in CC-Case. In other words, as a logical model, a process that can originally monitor the specific content of each process that is changing after the process that the component should have completed, is stored in advance. The technology is used for each process, but they are not intended to be used in the waterfall method, but are intended to be used for agile. We describe the specific usage of these technologies in each process in section 4.4 onwards.

4.4. CC-Case Requirements Analysis

CC-Case carries out requirements analysis from the planning and concept stage by using STAMP, an accident model based on system theory, in an expanded manner. However, the current security analysis method of STAMP has the problems introduced in Section 5.3, and the authors have proposed an integrated risk analysis method for safety and security based on system theory that can solve this problem. By proceeding with the control structure, hazard analysis and threat analysis are performed on the same control structure after understanding the overall picture.

4.4.1. What is STAMP S & S?

The authors call STAMP S & S the extended use of these analysis methods of STAMP beyond integrating safety and security. S & S is an abbreviation for Safety, Security, System, Software, Service, Society, etc. and means that these different viewpoints and different objects are analyzed for their interaction. In this case, "A" means not only the meaning of the accident indicated by the proponent, but also the meaning of the architecture considered in a wider range. Originally, STAMP deals with some security, but most of it has been analyzed specifically for safety. However, STAMP S & S considers not only safety but also various characteristics, such as cyber security, quality, and maintainability. As demonstrated, it can be analyzed for various objects. S & S refers to abbreviations such as System, Software, Service, Society, and Specification besides Safety and Security. The analysis of interaction is carried out for these different viewpoints and different objects.

The reason for writing this abbreviation is to draw out the potential of STAMP based on the wide range of applications of STAMP that can analyze the interaction of not only various devices and systems but also various components such as people and organizations. This is because STAMP S & S is based on CC-Case, which is a new framework aimed at establishing specific application methods without being limited to safety.

4.4.2. Why do Requirements Analysis with STAMP S & S?

STAMP is a top-down accident model that illustrates the relationship between what is supposed to be controlled from a bird's-eye view through system thinking. STAMP S & S uses this as an architectural model based on systems engineering, and the control structure (CS) created by this model is based on a common STAMP method. In CS, not only physical devices and computer systems but also software and people are analyzed as components. It analyzes human interaction with organizations, systems, hardware, and software.

In addition, the analysis of humans, including ML based on artificial intelligence, naturally falls within targeting humans. This is because artificial intelligence is a substitute for one of the human functions. In addition, as a social technology system, it can be applied from a high-layer analysis of risk, an analysis between organizations to a structural analysis of people, systems, services, and equipment. In any case, the system can be hierarchized and analyzed based on the concept of SoS, with the goal of preventing risks, i.e., damage, as well as analysis of the system construction goal itself at the top. As such, it can be used to find a law that works.

Real security and safety are achieved by analyzing requirements for complex systems from different viewpoints and from different viewpoints that have not been called safety in the past. The authors believe that by making it a common architecture, it can be widely implemented in society.

In the future, STAMP S & S will expand the scope of application and characteristics based on STPA as a requirement analysis to identify unsafe conditions in interactions corresponding too many characteristics, and CAST as a technique for analyzing the results of accidents that occurred. Evaluation will be carried out sequentially. This paper presents a framework for safety and security integration and shows its effectiveness, so please refer to section 5.1 for details.

4.5. Functional Design of CC-Case

CC-Case connects the results of STAMP safety and security requirements analysis to functional design. It is desirable that this functional design is designed like each function shown in IoT high reliability function. However, when it is unclear what functions are required, it is conceivable to use the FRAM functional resonance method to identify the relationship between functions.

4.5.1. Why Resilience Engineering (FRAM)?

FRAM identifies the relationship between hidden functions by simulating the relationship between functions. In addition, it was announced that not only the relationship between safety functions as in the past but also four security enhancement capabilities should be enhanced. The four security enhancement abilities are Monitor: Ability to detect dangerous signs, Respond: Ability to react quickly to signs, Learn: Ability to learn from past successes and failures, Anticipate: Ability to predict future risks is there. Instead of creating a perfect barrier, it has been proposed to realize security as a resilient security. The author expects that these applications will promote practical safety security design.

In addition, functional design can be made into a component that can be used for secure functional design simply by changing parameters by formalizing it, as in the Common Criteria Protection Profile (PP). In addition, the formalization by PP makes it easy for third parties to understand the contents of the function and to check the validity. Please refer to section 5.2 for details of this PP.

Moreover, SARM can be used to analyze comprehensive relationships between functions. SARM analyzes the interaction between actors in tabular form, but it can be used not only for actors but also for functions versus functions. It is also suitable for a comprehensive analysis of not only security and general functions but also other characteristics such as safety and security. For details of SARM, see Section 5.1.

4.6. CC-Case Software Programming

In the CC-Case, by using scenario functions as software programs, we realize the solution to the important safety and security issues of bugless and virus neutralization. Because it is a methodology that solves the crisis of software programs that are becoming increasingly complex and disorderly, it uses scenario functions. This full-scale implementation will be in the future.

Although it is possible to convert the entire system program into a scenario function, we see it appropriate to start with using it as a component in a mission-critical part of a device or system, or when using it to verify the relationship of the program data structure. In addition, application to artificial intelligence that takes advantage of the bugless state of the scenario function and virus neutralization is also a future goal. Negoro has already applied to Go rules and wants to verify the effects.

Also, we would like to consider an application method using artificial intelligence algorithms based on scenario functions in SARM, which is a method for extracting functional relationships in tabular form proposed by Kaneko. We believe that these will be useful for analyzing the relationship between a large amount of input information such as sensor information in automatic driving.

4.7. Assurance of CC-Case

The procedures of CC-Case have a dual-layer. The upper layer is named *a logical model*. The under layer is named *a concrete model*. The logical model and concrete model are shown in Figure10. The logical model shows the process structure developed in detail as much as possible, independent of a specific system. The logical model has a life-cycle process for each stage's process. The concrete model contains real cases corresponding to the specific system. The concrete model is decomposed logically until it describes evidence at the bottom layer. The concrete model remains evidence as a real case with approval results of customers. This evidence recorded in sequence can be used for verification. Risks may change frequently. It is necessary to keep evidence depending on changes. CC-Case supports changes by storing all evidence in a DB.

CC-Case has a two-layer structure consisting of a logical model and a concrete model (Figure 10). The logical model presents the process of creating an assurance case logically, and the concrete model describes the actual case. A logical model is an assurance case of a process that defines a procedure. The logical model consists of life cycle processes and processes at each stage. The concrete model is the assurance case of the deliverable, according to the actual case created under the lowest goal of the logical model. The concrete model is described by logical decomposition as appropriate until the evidence is presented at the bottom layer. The concrete model leaves the customer's approval result by the evidence in the actual case and the agreement as evidence. All of the evidence accumulates one after another, and as a result, can be used for argument. Requests are not deterministic and can change, but it is necessary to leave evidence in response to changes. Therefore, in CC-Case, all evidence is stored in the DB so that change requests can be met as needed.

For CC-Case assurance, GSN, which is a typical method of assurance cases, is used. Assurance cases are used for requirements extraction as well as testing and validation.

The reason why GSN is used is that it is suitable for notation verification/validation that explains the logical connection between assertion and rationale based on goals. Based on the case of the smart house used for IoT validation, please refer to the evaluation experiment results in section 5.5 for the effect.

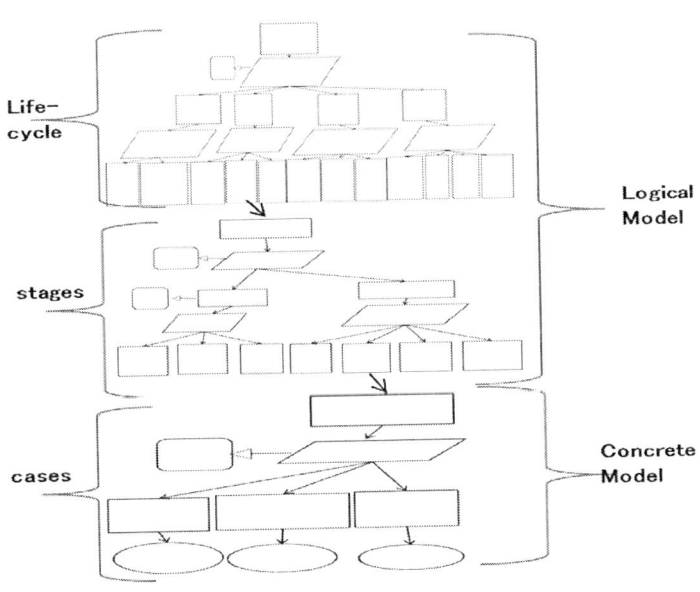

Figure 10. Logical model and Concrete model.

4.8. CC-Case Maintenance and Operation

CC-Case analyzes safety accidents and security incidents based on system thinking. For operation, the accident analysis method CAST is applied to safety and security.

As for maintenance, a scenario function is used to detect a defect in a program, and safe and secure software can be provided.

5. CC-CASE TECHNOLOGY ELEMENTS AND EXAMPLES

Among the technical elements included in CC-Case, some methods actually presented in the paper by the first author are explained as specific examples in this section. These correspond to the four items on the left, listed in the research history as CC-Case's Functions in the roadmap at the end of the book.

As an example of a cyber security response, 5.1 SARM comprehensive analysis of functional relationships is shown. In addition, 5.2 GSN quality assurance is shown. This is a CC-Case written in GSN using Common criteria as a process.

In addition, 5.3 STPA safety and security integrated analysis as an example of an integrated analysis of CPS safety and security based on STAMP.

At the end of this section, based on the characteristics of these studies, we examine the direction of future expansion.

5.1. Actor Relationship Matrix (SARM)

Actor Relationship Matrix (ARM) is a descriptive method which is able to generate a SD diagram, and increase understanding of the i* that is a representative goal-oriented approach. As an extension of the ARM, I propose a security requirements analysis method based on the Actor Relationship Matrix (SARM).

5.1.1. Purpose of SARM

The proposed method which analyzes the relationship between attacks and normal system functions is able to improve requirements analysis at system development, and then realize secure system development by considering attackers for actors.

5.1.2. Aim of SARM

The ARM has a feature that is better than a past requirements analysis technique such as i* as previously stated. But ARM doesn't become security correspondence. We propose 'Security requirements analysis technique (SARM) based on the ARM' that applies ARM to the security requirements analysis.

The aim of SARM is shown below:

1. Security requirements analysis which has complete coverage can be established by adding the attacker's existence to the ARM.
2. Developing a secure system can be established by examining the combination of the attack scene and the asset that is necessary to protect. There is an already-known pattern in the attack to some degree, by applying the attacking pattern of a web system such as STRIDE, fixed form analysis becomes possible.
3. Because expertise is needed for security, a general system developer takes charge of a general requirements analysis process and the design process. As for the threat analysis and the security function analysis, the method that the specialist separately executes is realistic. Then, the procedure can be analyzed dividing a function and a usual attack on a constant style.

5.1.3. Characteristic of SARM

This method is also able to convert ARM into SD diagram of i* of the Liu method for security, and has the following benefits: 1) Easy to create because of tabular form. 2) Easy to improve the integrity among related actors by verifying the relationship in tabular form. 3) Easy to express logically the relationship between tasks. I propose a spiral review process based on SARM and i* framework of Liu method as a combination of these methods as well.

5.1.4. Object Person and Advantage

SARM is assumed to be made by the developers to dig up specification of requirements. The advantage of SARM is that the security requirements analysis can be established by expressing dependency between actors including the attacker in each attack scene in table form with high convenience.

5.1.5. Example of Making SARM

In SARM, (1) AA (Asset and Attack) table (Table 9) (2) The SARM_A table (Table 10) exists.

Table 9. Example of AA (Asset and Attack) table

Method of attack	Type	Cookie	Session ID	Pass word	Individual inform-ation	SARM_A	SARM_B	SARM_C	SARM_D
Illegal order by identity theft	S	O	O	O	O	A_S	B_S	C_S	D_S
Falsification of ordering information	T		O	O	O	A_T	B_T	C_T	D_T
Negation to commodity order	R				O	A_R	B_R	C_R	D_R
Leakage of information	I	O	O	O	O	A_I	B_I	C_I	D_I
Dos attack to system etc.	D	O	O			A_D	B_D	C_D	D_D
Authority promotion to manager	E	O				A_E	B_E	C_E	D_E

→ SARM unit of making

5.1.5.1. Method of Making AA (Asset and Attack) Table

The AA table is a table where the combination of the asset and the attack is arranged to specify the asset that should protect on each attack scene. To narrow it at the level used in the requirements analysis process, AA table is patterned on the unit of STRIDE. The STRIDE model which Microsoft defines is the six types of the threat that are spoofing, tampering, repudiation, information disclosure, denial of service, elevation of

CC-Case 43

privilege. By making AA table on the unit of STRIDE, the security requirements analysis doesn't leak, and is not too detailed.

5.1.5.2. Method of Making SARM_A (Attack) Table

SARM_A table is an ARM that added attacker to general actor by each attack scene extracted in the AA table. A goal, soft goals, tasks, and the resources of a general person and the attacker are filled with structured expression.

- Each mark like ○which means goal, ☆ which means soft goal, ◇ which means task, and □ which means resource is filled in first before text starts.
- A general person must use the mark of white pulling out. The attacker must use the mark of black paint, to clarify attacker's target, intention, and task. The attacker's column is painted gray.
- See the Table 10. Example of SARM_A (Attack) table.
- The intention of "system BOB with the vulnerability" of the attacker Eve is expressed in the row of "system BOB with the vulnerability" of the line of the attacker Eve. Under the intention that "★Eve wants to order the commodity by disguising as ALICE in the system of BOB, and to obtain the commodity", the task of "◆ Session ID of ALICE is stolen." and the tasks of "◆ Eve disguises as ALICE, and the commodity is ordered", are executed together. So &, which means logical and are filled in first before ◆. As a resource that should be protected from the task of "◆ Session ID of ALICE is stolen.", there is ■-- COOKIE of Alice. The resource has the vulnerability of the second level. The vulnerability is displayed at three stage level of the $--,-,$ blank for the attacker's goal and soft goal. The level is based on the vulnerability of the Liu method.
- The task of "◆Session ID of ALICE is stolen." It can be done by either method of attacks which are the task of "◆Script Insertion"

or the task of "◆HTTP response attack" or the task of "◆XSS", so | which means logical or is filled in first.

- Moreover, "system BOB with the vulnerability" has the vulnerability of XSS. A special situation of XSS which it assists to the attacker's attack without intending can be expressed in the opposite angle column of the BOB line and the BOB row, and this part is painted out by the gray.
- Equivalence of both methods.

Table 10 in SARM, each item is numbered from S1 to S37. The item of SARM can be filled in matching every item of the Liu method of i* of Figure 11. Moreover, it is possible to fill it in matching everything from the Liu method of i* to the item of SARM. Then, the Liu method of i* and SARM have equivalence. If the conversion tool of i* and SARM is made, one of expressions can be converted into the other side easily.

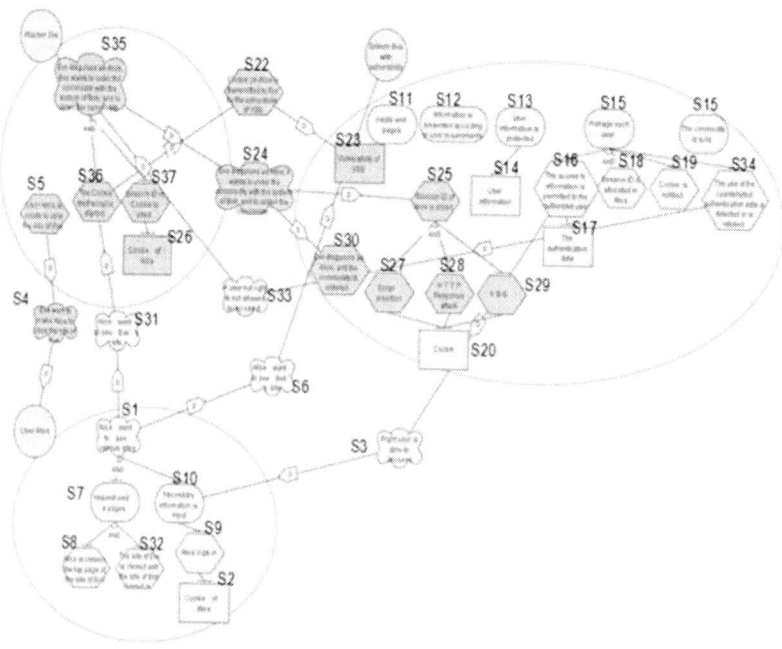

Figure 11. Example of the Liu method of i* matched to S1 to S37 of SARM.

Table 10. Example of SARM_A (Attack) table numbered from S1 to S37

	ID	User Alice	ID	System Bob with vulnerability	ID	Attacker Eye
User Alice	S1 S2	☆Alice want to see various sites. □Cookie of Alice	S6 S7 S8 S9 S10	☆Alice want to see Bob's site. ○ request web's pages. &◇Alice accesses the top page of the site of &◇Alice logs in. ○ Necessary information is input	S31 S32	☆Alice want to see Eye's site. ◇The site of Eve is clicked with the site of Bob logged
System Bob with vulnerability	S3	☆ Right user is made to access it.	S11 S12 S13 S14 S15 S16 S17 S18 S19 S20 S21	& O create web's pages. & O Information is presented according to user requirements. & O User information is protected. □User information & O manage each user. &◇ The access to information is permitted to the authorized user □ The authentication data &◇Session ID is allocated in Alice, &◇ Cookie is notified. ○ — Cookie & O The commodity is sold.	S33 S34	☆A user not right is not allowed to access it. ◇The use of the counterfeited authentication data is detected or is refused.
			S22	¦◆Cookie on Alice is transmitted to Eve by the vulnerability of XSS.		
			S23	—Vulnerability of XSS		
Attacker Eye	S4 S5	★ Eve want to click the site of Eve to Alice. ◆User Alice is made to click the site of Eve.	S24 S25 S26 S27 S28 S29 S30	★ Eve disguises as Alice, Eve wants to order the commodity with the system of Bob, and to obtain the commodity. &◆Session ID of Alice is stolen. ——Cookie ¦◆Script Insertion ¦◆HTTP Response attack ¦◆XSS &◆Eye disguises as Alice, and the commodity is ordered.	S35 S36 S37	★ Eve disguises as Alice, it wants to order the commodity with the system of Bob, and to obtain the commodity. &◆The Cookie theft script is started. &◆Session ID in Cookie is used.

5.2. GSN Quality Assurance with CC

5.2.1. Life-Cycle Support of CC-Case Life-Cycle Support of CC-Case

We show the support of the life-cycle process of CC-Case. The life-cycle process of CC-Case contains whole processes of requirement, design, implementation, test, and maintenance stages. However, in this research, we focus on the process of the requirement stage.

The life-cycle process of CC-Case should handle whole risks of security, including business continuity risk of security.

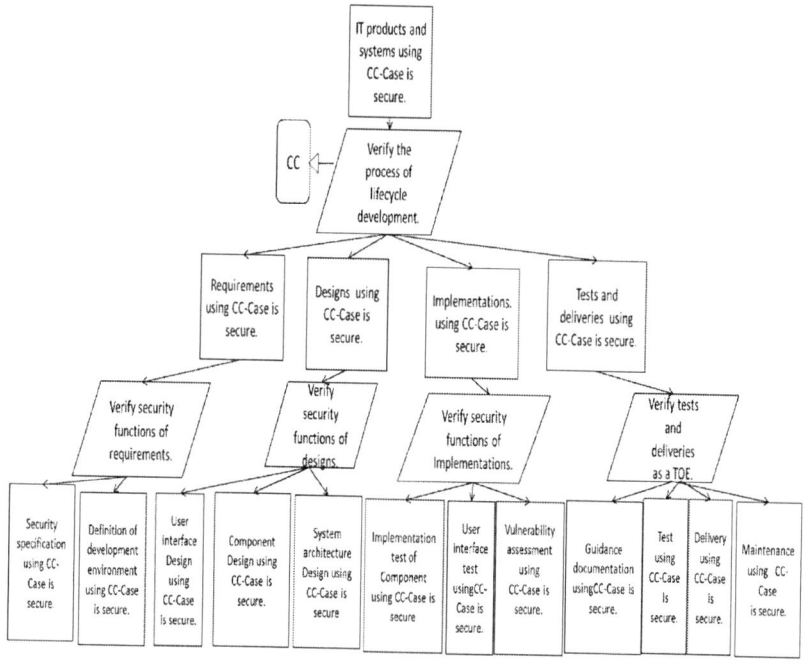

Figure 12. The life-cycle process of CC-Case.

Figure 12 is the life-cycle process of CC-Case. CC-Case uses GSN which is one common notation of assurance case. Using this assurance case, we explain the concept of the life-cycle process of CC-Case. In this case, top goal of assurance case is "IT products and systems using CC-Case are secure." Explicit assumption of the argumentations is "CC." The strategy shows how to verify the process of life-cycle development.

The strategy can be divided into 4 processes of the second goal, "Requirements using CC-Case is secure.", "Designs using CC-Case is secure.", "Implementations using CC-Case is secure," and "Tests and deliveries using CC-Case is secure." These goals need evidence which can verify the goals.

The second goal of "Requirements using CC-Case is secure" can be divided into 2 processes of the third goal, "Security specification using CC-Case is secure," and "Definition of development environment using CC-Case is secure" through the strategy of "Verify security functions of requirements." The third goal, "Security specification using CC-Case is secure" is equivalent to the top goal of CC-Case at the requirement stage.

5.2.2. The Requirement Stage of CC-Case

5.2.2.1. Assurance Case of Security Specification

In the requirement stage, the procedures to make security specifications are defined, and the documents which are necessary for ST (Security Target) are made. These procedures are defined as an assurance case, and produce evidence which give grounds of conformity with CC and agreements with customers. The assurance case of security specifications can be classified into the stage of defining security concept, of making measures and of making security specifications. Each stage's logical model is shown. Another purpose, player confirmation method for the output, and process (input, procedure, output) are clarified for every goal at the bottom layer of each stage.

Figure 13 shows the relationship of the procedures to make secure specifications, inputs, and evidence which give grounds.

5.2.3. The Merits of CC-Case by Using Life-Cycle Process by GSN

CC-Case has many merits to solve several problems which we face at the development of secure systems. In this paper, we show the merits focus on its life-cycle process. The life-cycle process of CC-Case can be expected to establish the discipline and control in the processes of refinement of the IT products and systems during its development and maintenance. It strengthens the handling for system risk and business continuity risk by life-cycle support.

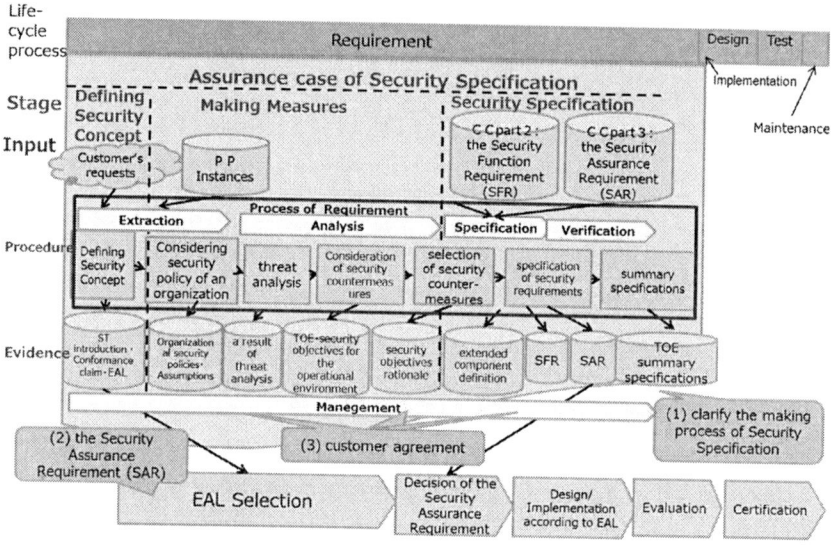

Figure 13. Whole model of the requirement stage of CC-Case.

The merits of handling system risk are mentioned below. CC-Case makes it possible to improve the development method to handle. The life-cycle process using CC-Case establishes rules and control in the development and assurance of objective system and production by implementing its requirements to system and production correctly.

For example, at the design stage, CC-Case makes it easier to accept specification changes by using its logical traceability and evidence. It can expect the improvement of development method, reuse, productivity that is the problem of the assurance case.

By defining development processes at each stage, CC-Case would be improving the development method with assurance of life-cycle. It can keep assurance based on CC through the life-cycle process. If its scope was only the requirement stage, it has only assurance at the time as the mere expectation. If its scope is the whole life-cycle process, it is possible to assure the real products of long span. In other words, CC-Case with life-cycle support has a different quality of assurance from CC-Case with requirement stage.

The merits of handling business continuity risk are mentioned below. Security risks change incessantly because invisible attackers exist, and an unexpected new threat occurs. The life-cycle support of requirement, design, test, and maintenance makes countermeasure easily against the situation which an unexpected new threat produced. The life-cycle process of CC-Case can handle the business continuity risk by its monitoring and control process. It makes it easier to handle additional changes of risk.

If you make a system or a production with CC-Case, it makes it easy to cope with modification for a security accident. That is because CC-Case has evidence of the argumentations to clarify a modification point for an essential cause to a security accident. At the maintenance stage, CC-Case can be expected to improve reusability, productivity by reusing evidence stored according to defined process.

5.3. Safety & Security Integrated Analysis by STAMP/STPA

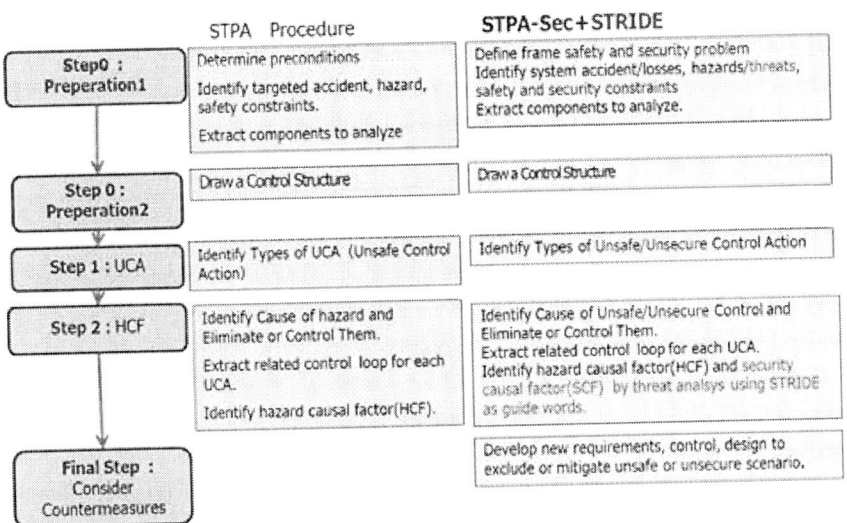

Figure 14. Process of STPA-Sec + STRIDE method.

We propose the procedure for applying STRIDE to STPA as follows [54]: The difference between STPA and STPA-Sec is shown in blue letters and the part added by STRIDE application is shown in red letters (Figure 14).

5.3.1. Safety and Security Framework Setting

STPA-Sec introduces a framework that includes hierarchy. Six categories of STRIDE are additionally applied as hint words of security causal analysis in Step 2. Safety and security are handled in parallel on the same Control Structure.

5.3.2. Security by Design

Incidentally, although there is a detailed part added in red in Step 2 in the STPA-Sec procedure, our + STRIDE method does not adopt the procedure. Step 2 of STPA identifies the SCF for the unsecure control action identified in Step 1, but the further detailed procedure is the range of tailoring. The part added in Step 2 of STPA-Sec appears to be effective in problem analysis in the concept phase, but it is not applicable because it is targeted for security design in the + STRIDE method.

5.3.3. Modeling against Threats

As a hint word of security factor analysis in Step 2, threat analysis such as that provided by STRIDE can be added to the threat modeling. Step 2 is a state after what was a security problem in the previous steps. We identified the assets to be protected as components of the control structure. It is the state after the control action, which is that the interaction between the components is unsecure. Identifying the unsecure state using the four guide words does not reveal the vulnerability information of the device itself but does identify the unknown threat. We will identify the security factors for this unknown unsecured state (= threat). The proposed method using STRIDE for the first time in Step 2 makes it easy to distinguish between factors and countermeasures at this stage and details the procedure of threat modeling that STPA does not specify in detail.

In other words, the proposed method includes comprehensive identification of the security risk as an interaction advocated by STPA-Sec.

5.3.4. Confidentiality

Using STRIDE enables analysis based on attributes such as confidentiality rather than availability or integrity.

5.3.5. Microgrid Examples of STPA-SafeSec

The document [55] uses a microgrid as a case study, with the analysis of the connection between a wide-area power network and a local power network.

The procedures for security are considered according to the contents of the case.

- **Step 0 Preparation 1** (STPA-SafeSec II ~ IV) identifies safety constraints and security constraints for each hazard. We identify the safety and security constraints at a high level of abstraction by taking the negative form of the hazard with the constraints being safety constraints (Safety, CSTR-S-n) and availability constraints (Availability, CSTR-A-n). We number the constraints according to their attributes, such as integrity constraints (Integrity, CSTR -I-n). In this case, only CSTR-S-5 from the safety constraint CSTR-S-1 (the negative form of H1 to H5) appears, but the availability and integrity constraints are generally handled.
- **Step 0 Preparation 2** (STPA-SafeSec V) builds with the control structure in the control layer.
- **Step 1** Extract UCA (STPA-SafeSec VI ~ IX).
- **Step 2a** Build a component layer (STPA-SafeSec X, XI). At the physical level, the speed controller at the functional level is represented by a specific configuration such as analog digital converters and Raspberry Pi, with USB.
- **Step 2b** (STPA-SafeSec XII) allocate abstract safety and security constraints to component layer elements.

- **Step 2c** (STPA-SafeSec XIII) detailed the abstract hazard scenario to the component layer.

    ＊For easy understanding, it is divided into Step2a, b and c.

The STPA-SafeSec paper considers the procedure to be divided into the control and component layers. The specific case of STRIDE applied to the control layer is shown in Table 14. It applied to the component layer is shown in Table 15.

5.3.6. Effect of Applying STRIDE on the Control Layer

The principal aim of this paper is to demonstrate that security analysis is possible using a threat modeling by applying STRIDE even in a control layer on which STPA-SafeSec cannot analyze security.

For this reason, we specifically apply STRIDE to an additional part of the threat analysis that is not specified in the STPA-SafeSec and STPA-Sec processes. As a procedure, the SCF analysis of Step 2 will be added. This shows the flow when it is implemented using STPA-Sec and + STRIDE.

- **Step 0 Preparation 1:** System Engineering Foundations

Define the frame safety and security problem

Assure a safe and secure power operation. Today, threats of cyberattacks are increasing in power grid operation, including maintenance. There can be many attack elements including terrorism.

Identify losses or accidents following A1-A4 in Table 11. STPA-SafeSec can also be used in the same manner.

Next, identify hazards (H1-H7) and threats (T1-T3) in Table 12. In our opinion, not only hazards but also threats must be identified in this step because safety hazards represent security threats.

Draw a control structure of the control layer. Figure 15 shows the example of microgrid used in the STPA-SafeSec paper.

Table 11. Identify losses or accidents

ID	Accidents/Losses
A1	Injury to humans
A2	Damage to power equipment
A3	Damage to end-user equipment
A4	Interruption of power supply to consumer loads

Table 12. Identify hazards (H1-H7) and threats (T1-T3)

ID	Hazard
H-1	Out-of-sync reclosure
H-2	Operation of power equipment outside of operational limits
H-3	Violation of power quality metrics
H-4	Inability to achieve synchronization
H-5	Inability to meet local demand
	Threat
T-1	Power equipment is destroyed
T-2	Operation of power equipment is deprived of authority
T-3	Control information of device is stolen

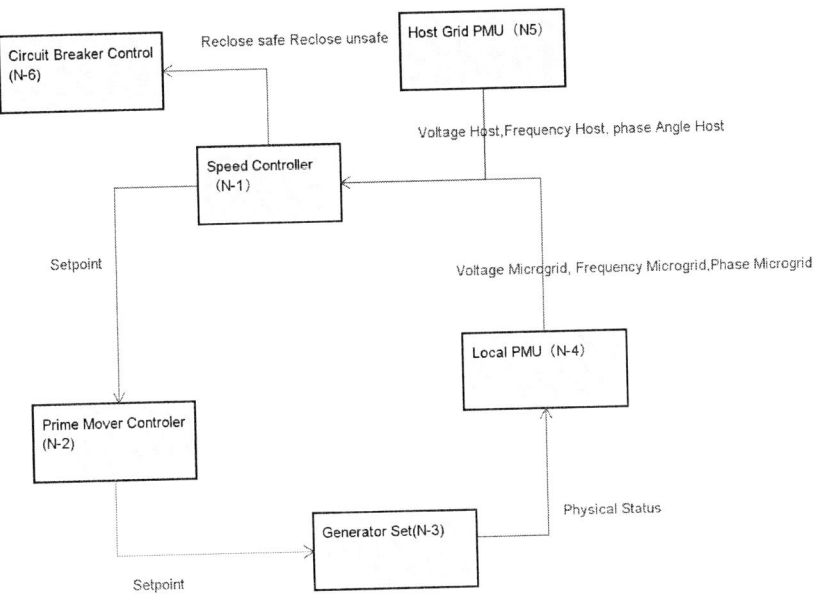

Figure 15. Control structure of Control layer.

Table 13. Unsafe or unsecure control action

No	CA	From	To	Not Providing	Providing causes hazard	Too early or Too late	Stopping too soon or applying too long
1	Reclose safe Reclose unsafe	Speed Controller (N-1)	Circuit Breaker Control(N-6)		The speed controller wrongfully assumes that synchronization is achieved. It would then indicate that the reclosure of the circuit breaker is safe when it is not. (H1)	The speed controller assumes that synchronization is achieved. It would then indicate that the reclosure of the circuit breaker is safe while it is too early or too late.(H1)	
2	Setpoint	Speed Controller (N-1)	Prime Mover Controller (N-2)	When the breaker is in the released state, set values within the operating range are instructed to the prime mover controller with Not (In other words, the setting value is not updated)(H-3,H-4,H-5)	Instructs the prime mover controller to set the value outside the operation range(H-2,T-1,T-2)	When the breaker is in the released state, set values within the operating range are sent as instructions to the prime mover controller with Too late. (In other words, the setting value is not updated) (H-3,H-4,H-5)	
3	Voltage Host, Frequency Host, Phase Angle Host	Host Grid PMU (N5)	Speed Controller (N-1)	Host Grid PMU does not report measured voltage Host, Frequency Host, or Phase angle Host. (H-3,T-3)	Host Grid PMU reports incorrect measured voltage Host, Frequency Host, or Phase angle Host. (H-3,T-1)		
4	Voltage Microgrid, Frequency Microgrid, Phase Microgrid	Local PMU (N-4)	Speed Controller (N-1)	Host Grid PMU do not report measured voltage Host, Frequency Host, Phase angle Host. (H-3,T-3)	Local PMU reports incorrectly measured voltage Host, Frequency Host, or Phase angle Host. (H-3,T-1)		

* PMU: Phasor Measurement Units

CC-Case 55

- **Step 1:** Identify Unsafe or Unsecure Control Actions

Unsafe or Unsecure control action from speed controller is shown in Figure 15 in red. In this case, instructing the prime mover controller to set the value outside the operation range not only leads to hazard (H-2) but also threatens (T-1, T-2). Hence, this control action is unsafe and unsecure. Table 13. shows 4 types of unsafe and unsecure status for each control action (CA).

- **Step 2:** Identify causes of unsafe or unsecure control and eliminate or control them according to STPA-SafeSec [49], security constraints are identified in the control layer, but a security analysis is not performed.

However, if we follow the principle of Security by Design, a security analysis for the speed control (N1) of the control layer is necessary even in the early stages of development. We also implemented STRIDE analysis, a threat analysis method that can be used in the early phase. In other words, a threat analysis based on an attacker's perspective was conducted to derive the hazard scenario.

Specifically, we attempted to analyze the speed controller in the control layer using STRIDE. STRIDE is based on a reference architecture for determining the overall image of a system. Enumerating threats as a threat analysis diagram is used to verify mitigation and mitigation measures. The purpose of threat modeling, for which it would be possible to use a layer, was to understand how an attacker could penetrate the system.

It is important to take appropriate mitigation measures and to consider mitigation measures in the early design phase rather than after the system is deployed to eliminate the cost waste. Therefore, in Step 2, we use STRIDE to analyze each attribute and analyze the SCF to ensure that security functions are held as identifiable measures.

In this case study, we focus on the speed controller against the unsafe or unsecure CA of "instructs the prime mover controller to set the value outside the operation range (H-2, T-1, T-2)." STPA-Sec extracts causal factors by adding guide words such as "malformed" or "unauthorized." However, in this study, we decided to use STRIDE as a guide word for this SCF.

The controller corresponds to the speed controller in this case. The hint word of STRIDE will be used to refine the process model and control the algorithm of the control input, malformed external information, malformed feedback, or the controller itself, which is the input of the following item controller. To identify the hazard factor as the hint word of STPA instead of STPA-Sec, the hint word specified in STPA is used as it is.

Table 14 shows the SCF for the speed controller (CPU). In this case, we extracted STRIDE as a hint word in the threat scenario and took sample countermeasures against it. The six classifications of STRIDE represent the required security properties of authentication, integrity, confidentiality, availability, and authorization, as shown in Table 14, and the threats listed from different perspectives show the direction of mitigation for each causal factor. Using a classification Threat Tree [56], which shows the threat mechanism linked to STRIDE, the threat reduction by development and operation is obtained for each threat identified to the point where the attack can occur in STPA. STRIDE has traditionally been used for threat modeling of information systems, but its application to IoT security, including devices that are connected with special applications such as these, has been discussed. Hence, while referring to this commentary in the case of SafeSec, "scenario 1.1: CPT-N1 Speed Controller" recognizes the correct feedback incorrectly. We analyze the SCF in STRIDE. The SCF for the speed controller (CPU) of this case is shown in Table 14 which was extracted from STRIDE as a hint word.

"Repudiation" does not have a corresponding factor in N1-1 because the user has no means of proving this action. In addition, in Table 12, we show the specific threat scenarios and countermeasures for each SCF of N1-1, based on the STRIDE analysis of IoT security [17]. In this analysis, it becomes clear that the attacker's perspective is capable of any kind of attack on the target device. Moreover, each threat is classified by a security property, and hence it is easy to determine the necessary security measures.

In addition to the analysis for devices called N1-1, it is possible to guide threat scenarios and countermeasures according to the target, such as communication and storage.

Table 14. SCF of N1, Threat Scenarios, and Countermeasures for the Speed Controller (in the Control Layer)

STRIDE	Required Properties	SCF of N1	Expected threat scenarios	Example of measures
Spoofing identity	Authentication	No correct authentication is made for N1-1 (Speed controller) (N1-S)	Host PMU impersonates the local PMU	Use IC chip with authentication function
Tampering	Integrity	Incorrect FB signal is inserted into N1 (Speed controller) (N1-T)	Some or all of the software running on the speed control is replaced by an attacker	Message authentication Code (MAC), tamper-proof mechanism applied to speed controller
Information Disclosure	Confidentiality	The FB signal of N1-1 (Speed controller) is leaked (N1-I)	If the software running on the speed controller has been modified, the modified software might disclose the plaintext to an unauthorized person.	Implemented anti-malware, Secure Key Management
Denial of Service	Availability	N1 (Speed controller) is destroyed (N1-D)	• The speed controller is exposed to the threat of DoS in the form of constantly waiting for the network for incoming and unsolicited datagrams. • An attacker can open a large number of connections at the same time and take an extremely long time to process. In some cases, one-sided traffic can undermine the speed controller's ability to handle it. In both cases, the speed controller is virtually a malfunction in the network. -The function of the speed controller stops or cannot communicate by interference or cable cutting.	Limit the number of accesses from an attacker or the same IP. Create a speed controller that can withstand large-scale traffic
Elevation of Privilege	Authorization	(N1-E)	Limit the number of accesses from an attacker or the same IP. Create a speed controller that can withstand large-scale traffic	Access control of the speed controller. Establish an authorization scheme.

SafeSec analyzes vulnerabilities that could be security threats, but provides no threat analysis from the attacker's perspective, and only availability and integrity are considered as security properties, excluding access control, confidentiality and authentication that results in encryption. This can be accomplished by combining STPA-Sec with STRIDE analysis, which is considered to be an analysis of authenticity that leads to authorization. For the speed controller, one of the components of the control layer, the specific threat scenarios and countermeasures that are assumed for each property could be addressed as a result of the STRIDE analysis shown in Table 14.

5.3.7. Effect of Applying STRIDE on a Component Layer

We discussed how the choice of component layer can be affected by the STRIDE application of the control layer speed controller. Raspberry Pi is one of the components of the component layer. As a result of the STRIDE analysis shown in Table 15, we were able to extract the specific threat scenarios and countermeasures for Raspberry Pi that were expected for each property.

Table 16 lists the evaluation by comparison from the perspective to be considered when comparing the security analyses of STPA-Sec, STPA-SafeSec, and STPA-Sec (+ STRIDE). The results are assessed based on the requirements of the (1)-(3), and sufficient conditions are shown in Table 16. Because different properties are required for the target system and the product, it is not possible to determine confidentiality as well as safety, availability, and integrity. Therefore, there is no evaluation based on the sufficient conditions required at this time. In summary, STPA-Sec and + STRIDE are more highly rated in terms of (1) safety and security framework setting and (2) Security by Design. (3) In modeling against threats, + STRIDE is more valuable than STPA-SafeSec and STPA-Sec.

In the case of STPA-SafeSec, no security analysis is performed on the speed controller in the control layer. However, we identified seven concrete SCF, scenarios, and countermeasures using STPA-Sec (+ STRIDE) in the control layer. In additon, STPA-Sec (+STRIDE) were able to identify six SCFs and more concrete security scenarios and countermeasures for Raspberry Pi in the component layer.

Table 15. SCF of N1-1, Expected scenario, and measures of Raspberry Pi (in the Component layer)

STRIDE	Required Properties	SCF of N1-1	Expected scenarios	Example of measures
Spoofing identity	Authentication	No correct authentication is made to N1-1 (Speed controller CPU) (N1-1-S)	If the operating system user settings are not set properly, attackers might spoof them	Set the password appropriately: SSH Login with private key
Tampering	Integrity	Incorrect FB signal is inserted into N1-1 (Speed controller CPU) (N1-1-T)	If an illegal program has access to a cryptographic key or an encryption mechanism that holds the cryptographic key, the software replaced will misuse the real ID of the speed controller. An attacker can use the extracted cryptographic keys to intercept, block, and replace data from the speed controller with false data and pass authentication with a stolen cryptographic key.	MAC Applying a tamper-proof mechanism to the speed controller
Repudiation	Accountability	(N1-1-R)	If the Raspberry Pi user does not have a log of the communication, it is likely to negate the fact of the operation that the user performed improperly	Acquisition and maintenance of various logs
Information Disclosure	Confidentiality	The FB signal of N1-1 (Speed controller CPU) is leaked (N1-1-I)	The attacker exploits the encrypted key and obtains the encryption key and decryption key between the speed controller and The Controller (the field gateway or the Cloud gateway), thereby allowing the attacker to get the clear text.	Implemented Anti-malware, Secure Key Management
Denial of Service	Availability	N1-1 (Speed controller CPU) is destroyed (N1-1-D)	The function might be stopped if unauthorized access is performed over a WAN or Ethernet, or when a large amount of data is received.	Apply response limit
Elevation of Privilege	Authorization	(N1-1-E)	If the administrator setting of the OS is not appropriate, the user who does not have administrator rights of the OS originally has administrator privileges, and execution with administrator authority might be used illegally	"Run as Administrator" or "Restrict users who can get administrator rights"

Table 16. Evaluation of STPA-Sec and STPA-SafeSec

Methods	(1) Safety and Security framework setting		(2) Security by Design		(3) Threat modeling	
Condition	necessary condition	sufficient condition	necessary condition	sufficient condition	necessary condition	sufficient condition
Evaluation criteria	(1) The ability to analyze safety and security together	(1) Safety and security are analyzed at any stage (coverage)	(2) Thinking about security from the top down. (Cost reduction)	(2) Obtaining necessary security requirements in the early stages and make it into security design. (validity)	(3) Be able to analyze risks based on vulnerabilities or threats	(3) Both vulnerabilities and threat-based risks have been extracted (logic)
STPA-SafeSec.	STPA-SafeSec derives system hazards and safety and security constraints, but does not perform security analysis on the Control layer.		It does not consider security at the top down from the early stages of planning and requirements definition processes.		The security analysis in the Component layer of STPA-SafeSec is conductor to analyze the well-known security vulnerabilities of physical equipment.	
Evaluation	Yes	No	No	No	Yes	No
STPA-Sec	STPA-Sec does not differentiate between the analysis stage of safety and security.		STPA-Sec has a top–down approach. This contributes to cost reduction. However, this is focused on problem analysis in the concept stage for judging at the business level, and no method for creating security requirements to turn into security functions is presented.		STPA-Sec contains the problem, vulnerability, and threat analysis, but the procedures and cases of systematic threat analysis have not been made public.	
Evaluation	Yes	Yes	Yes	No	Yes	Yes (However, details of threat analysis not presented)
STPA-Sec +STRIDE	This is the same as STPA-Sec.		This is a top–down approach that is the same as STPA-Sec. Furthermore, it is possible to present a method of creating security requests by security functions using STRIDE threat analysis.		STPA-Sec (+STRIDE) perform threat modeling in addition to STPA-Sec analysis.	
Evaluation	Yes	Yes	Yes	Yes	Yes	Yes

As selecting the component layer after applying STRIDE to the speed controller of the control layer is available, selecting a physical device that can create an appropriate security function that can take countermeasures against the threats derived by the STRIDE analysis is possible. For example, in that case, to take countermeasures against the threat of spoofing the speed controller, a physical device having a more advanced authentication function must be selected. In that case, there is a possibility that Raspberry Pi is not selected. A more secure and highly functional device will be selected.

6. CONSIDERATIONS FOR CC-CASE

We show the challenges to cybersecurity in Section 5, but this section briefly describes the significance of integrated use of CC-Case technologies and which technical elements of CC-Case have been applied as research to IoT and AI.

6.1. The Integrated Use of CC-Case Technologies

Up to the previous section, we have presented how to use CC-Case technology for each life cycle. Furthermore, in this section, we will consider not only the individual use for each process but also how to use each technology in an integrated way and present hypotheses. In order to use each technology in an integrated way, it is necessary to use the technology according to the characteristics of each technology element.

What are the features of STAMP, GSN, FRAM, and SARM that can be used to identify requirements?

- STAMP CS understands the overall structure. It is suitable for grasping the control structure of a physical system, but this can also be applied to the cyber area of software, people, and organization.
- GSN visualizes the state and grasps the logical structure. This is effective for visualizing invisible logic such as the cyber area of software. In addition, requirements can be guaranteed by formalizing into the limited logic of goals, assumptions, explanations, and evidences.

- FRAM visualizes the function, that is, the movement, and grasps the relationship. Unlike GSN visualizing the state, we focus on the movement itself.
- SARM performs a comprehensive analysis of the functions of safety and security (system movements) and actors (human movements).

The overall structure by STAMP, the state by GSN, the visualization of movement by FRAM, and the exhaustive analysis by SARM will result in the extraction of different requirements, so we think that integrated way will create value in realizing safety and security.

The concrete integration method is based on STAMP CS, which is a notation that shows the structure of the entire system, and processes and algorithms are used for the internal structure of each component. In other words, a safe and secure state is assured in the life cycle by creating and managing algorithms and processes for each component. The algorithm is software when the component is a normal system, but is a mental model in the case of a human, and a machine learning algorithm (ML) in the case of artificial intelligence (AI). It is necessary to visualize the state of these algorithms. This could be used to visualize ML black boxes. Assurance cases are assumed to be available as a means of extracting these requirements and verifying/validating them.

As the process of each component, we are considering using the assurance case with the process described by GSN (CC-Case in the narrow sense). In addition, the author is thinking of building a foundation that can manage that this assurance case keeps the components in the CS diagram kept safe and secure dynamically.

The identification of functions by FRAM may be effective when the relationship between functions is unknown and as a result it is impossible to predict what kind of hazard may occur. In the development of space systems, it is used to identify demands that make it difficult to find out what kind of danger exists when searching for Mars. In addition, since the analysis involves simulation, it will be possible to respond to requests to change constantly. The most appropriate use of this method is an important issue for truly resilient safety security. The author is considering using each component of the CS diagram as a function and using FRAM to identify the risk.

Examining the difference between STAMP/STPA and risk extraction is also considered as a proposal for finding the most suitable usage.

Because SARM is a technique to improve the completeness by converting from a diagram format to a tabular format, FRAM converts requirements from a graphical representation to a tabular representation, enabling comprehensive analysis. In addition, it is suitable for managing the relationship of a large amount of information such as sensor information.

The Scenario Function visualizes and makes sense of software data relationships. He is also a savior in the safety security of systems that include bugless and virus neutralization software. As for the scenario function, it is very revolutionary in itself, so I would like to try it out and use it to the fullest, based on the facts.

CC is a standard that can be called the culmination of IT security, in which PP formalizes design requirements. The author envisions extending this security requirement and assurance process to one with safety. I would also like to use the PP catalog as a way to formalize the design.

As mentioned above, the authors are considering STAMP S & S to use the Control Structure (CS) diagram of STAMP presented by Leveson as the basis for the structure of the entire system. The reason for this is that CS has a feature that makes it possible to describe all things in the world in a very concise and necessary way.

In order to accurately describe all things in the world, philosophy is needed as a way of understanding the world. Unlike Western philosophy, Eastern philosophy, especially Buddhism, tries to grasp life as the reality of all things. In Buddhism philosophy, factors common to all life [57] is divided into the following 10 categories, such as "the appearance, nature, entity, power, influence, internal cause, relation, latent effect, manifest effect, and their consistency from beginning to end." The CS diagram includes these all factors precisely.

In order to implement the system safely and securely, it is necessary to accurately grasp the factors of everything. The CS describes the interaction between the control component and the controlled component as a whole.

This component can be almost anything: equipment, system, people, organization, and so on. Components include algorithms and processes. In the case of human, the algorithm is regarded as a mental model. In terms of 10 factors common to all life, the component of the object is the entity, and it is divided into its appearance and nature.

In addition, each component is modeled to be exchanged by control action and feedback. In terms of control engineering, an actuator that amplifies force and a sensor that collects information are also added. In terms of 10 factors common to all life, the control action is regarded as power and the feedback as function.

Furthermore, STPA can analyze the causes, the hazards that triggered them, and the results.

STPA can be said to analyze the cause and relation (condition) and effect, which are the ties between them, for each power.

In addition, the manifest effect (reward) is considered an accident, or a goal to prevent the accident.

The CS diagram takes these components into systems thinking and captures that everything is connected and interacting with each other. That is a same idea as that "consistency from beginning to end" unifies factors.

The overall view is that a wide range of systems and devices are connected, like the IoT, from a highly abstract layer that broadly perceives society itself as one system, and the environment and people are also aware of one layer and AI. It can be handled hierarchically from one system and individual device layer, each component in the device, and the function in the software, and the relationship from the layer with high abstraction. It can be captured so that it can be tracked.

In this way, the great advantage of STAMP theory and its analytical methods is that the elements necessary for the formation of things and their relationships can be visualized in a very sophisticated and simple way.

By grasping the relationship by systems thinking that universally captures such a system, it is possible to comprehensively grasp where and what problems exist and what affects them. This helps reduce unexpected accidents, prevent incidents in advance, and reduce risk. It is necessary for the safety and security of the Internet-connected world to have an accurate map that captures the formation of such things using a systems approach. However, it is important not only to have a map, but also how to use it. CC-Case performs hazard analysis with STPA and extended threat analysis as described in the next section. Furthermore, it is very important not only to analyze risk but also to learn from actual accidents and connect them to proactive measures. CC-Case does this with CAST, an accident analysis method based on system theory.

In addition, as Hollnagel stated, after recognizing that an accident would occur, to take more resilient measures, it was much less probable than when an accident did not occur. So the approach to learning what works is very different from the traditional approach of identifying hazards and taking risk measures, but it is important. CC-Case not only thoroughly identifies risks to prevent accidents like STPA, but also analyzes with the goal of finding hidden laws that are working well like FRAM. This can be done based on the CS diagram.

CC-Case should be able to support complex systems in the AI/IoT era because it grasps all the elements in this way.

6.2. CC-Case Technologies Applied to IoT and AI

For IoT, we applied it to the validation of smart home threat extraction and conducted an evaluation experiment.

We have proposed CC-Case that is a security requirement analysis and assurance by using the Common Criteria (CC) and the assurance case. The assurance case is used by life-cycle in CC-Case. Therefore, it is the main factor of CC-Case. The assurance case of CC-Case is not the description of GSN but the method which defines the process as the logical model, and presents as the concrete model that conforms to the process. However, the effectiveness of CC-Case has not been shown. In this paper, we compared CC-Case which has a logical model of threat analysis process, GSN, and non-structured natural language representation by testing practice. We tested the effectiveness visualization and verification of the requirements of CC-Case.

GSN is intended to be formalized with contexts, goals, strategies, evidences, etc. but unless it is described according to the object, it will be mass production of irregular and difficult to understand GSN. That is the challenges of GSN. Therefore, the GSN of CC-Case is not simply written in GSN, but in the two-layered logic layer, a process corresponding to the target is defined and we confirm the validity. With security requirements, the process according to the target is the Common Criteria process as in the example presented in Section 5. In the threat extraction of smart homes, the threat analysis process was logically modeled regarding the Common Criteria process.

For AI, STAMP S & S was applied as a research experiment to level 3 risk analysis for automated driving. Safety and security were analyzed on one CS, and we conducted a comparison experiment to see how the risks differ when using STAMP S & S and when not using it.

(1) Safety and security can be analyzed from the planning/ requirements definition process. (2) As a safety analysis, we have extracted many hazards and factors related to human life and health. It is characterized by the ability to identify human errors, differences in human and system recognition, and system performance limits (SOTIF) that are not covered by functional safety standards and conventional safety analysis methods. SOTIF assumes safety risks other than system errors, such as driving other drivers and the weather. (3) As a security analysis, there were 33 threats extracted with STAMP S & S method using STRIDE as a keyword compared to 1 threat extraction with STPA-Sec keyword. In addition, when STAMP S & S was not used in the STAMP S & S method, only 3/22 people could raise the security risk, but all subjects raised the security risk. In this way, STAMP S & S showed good results.

In the future, we will also analyze the risk of input to AI such as image data and speed, and proceed with CC-Case application focusing on machine learning (ML) safety and security.

So, whether CC-Case can respond to the following four AI issues, consider:

(a) Ensuring safety system

Analyzing the entire system including ML by system thinking STAMP/STPA.

We can consider safety in STAMP S & S after modeling a complex system such as IoT as a whole system. At that time, machine learning (ML) is regarded as one of the components.

The author's paper [58] "The modeling pattern of human and society for AI business" shows human and social modeling patterns for AI business.

In this way, research that can use AI safely and securely in organizations and societies by structurally grasping the people and society that use AI, and analyzing new usage methods and hidden risks using STAMP S & S.

(b) Engineering framework to solve problems efficiently

Comparison between ML model (artificial intelligence) and safety model (natural intelligence)

There are two types of safety analysis: top-down type (STAMP) and bottom-up type (FRAM). By designing the architecture of the AI application system that integrates these methods, we aim to build an engineering framework that solves problems efficiently.

(c) Countermeasures to the black box problem

Select an ML algorithm that can assure safety.

Applying safety analysis to ML itself also realizes to make black boxed ML algorithm into a white box.

However, since ML is software, clarification of data connection by scenario functions will contribute.

(d) Quality assurance of machine learning itself

Analyze risks in each process of ML.

It can be achieved by securing 1, 2, and 3 as evidence. Specifically, an assurance case (GSN) a basis for dynamically managing evidence.

Furthermore, we are studying the creation of human and social modeling as a pattern by using AI [59]. We will study how AI as a substitute for humans can be used effectively from the perspective of social technology systems. One of the future plans is to clarify the problems in the implementation of AI through PS-Case [60] that systematizes the problem solving technique as an assurance case.

CONCLUSION

The trend of the times changes constantly. As shown in the roadmap (Figure 16), we have been developing a CC-Case development methodology based on the changing needs of cyber security, IoT, and AI, based on the needs of the trend.

In order to realize the needs of cyber security systems and security-by-design, we began by considering SARMs that comprehensively extract security risks. We considered the effectiveness of quality assurance by GSN and CC processes and quasi-formal methods of PP. In the IoT era, it became necessary to ensure safety that was not considered in conventional cybersecurity. IoT high-reliability function and GSN IoT applications were advanced. In addition, systems engineering for system thinking has become necessary for complex mission-critical systems. These risk analyses require safety that integrates safety and safety analysis based on safety theories such as STAMP and FRAM, leading to the proposal of STAMP S & S.

In the coming AI era, a CC-Case development method will solve the above four AI issues and will be evaluated through experiments. In particular, the scenario function with software visualization and virus neutralization, which is not an exaggeration to dominate the system is the key to safety and security integrity. CC-Case includes various technical elements, but we hope that a safe and secure system for a new era will be built by appropriately linking them, and CC-Case will become a trustworthy engineering method that goes beyond safety and security.

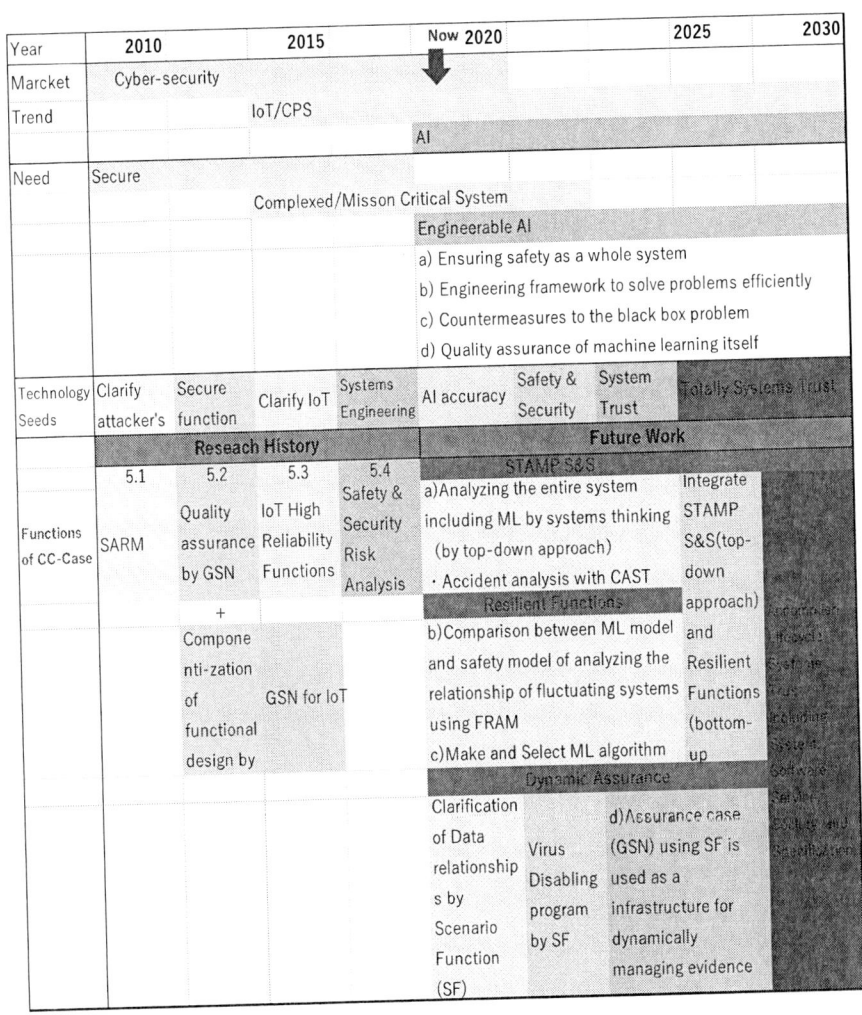

Figure 16. Roadmap of CC-Case. Trustworthy Engineering Methodology for AI/IoT.

REFERENCES

[1] IEC. *IEC 61025: 2006 Fault Tree Analysis (FTA)*. https://webstore.iec.ch/publication/4311.

[2] United States Military Procedure, *"Procedure for performing a failure mode effect and criticality analysis,"* November 9, 1949, MIL-P-1629.

[3] IEC 61882: 2001 *Hazard and operability studies (HAZOP studies) - Application guide.* http://www.iec.ch.

[4] Leveson, G. Nancy. *Engineering a Safer World*, MIT Press, 2012.

[5] *STPA handbook*, http://psas.scripts.mit.edu/home/.

[6] *CAST handbook*, http://psas.scripts.mit.edu/home/.

[7] Leveson, G. Nancy. *STAMP Intro*, http://sunnyday.mit.edu/workshop2019/STAMP-Intro2019.pdf.

[8] Hollnagel, E. *FRAM - The Functional Resonance Analysis Method: Modelling Complex Socio-technical Systems.* Farnham, UK: Ashgate., 2012.

[9] *The Functional Resonance Analysis Method*, http://www.functionalresonance.com/support/index.html.

[10] Hollnagel, E; Woods, D; Leveson, NC. (Eds.). *Resilience engineering: Concepts and precepts.* Aldershot, UK: Ashgate., 2006.

[11] Hollnagel, E. *Safety-I and Safety-II: The Past and Future of Safety Management.* Farnham, UK: Ashgate., 2014.

[12] *Resilience Engineering*, http://erikhollnagel.com/ideas/resilience-engineering.html.

[13] *Common Criteria for Information Technology Security Evaluation*, http://www.commoncriteriaportal.org/cc/.

[14] *US Pat. No. 10,235,522.*, http://appft.uspto.gov/netacgi/nph-Parser?Sect1=PTO2&Sect2=HITOFF&p=1&u=%2Fnetahtml%2FPTO%2Fsearch-adv.html&r=1&f=G&l=50&d=PG01& S1=15423481&OS=15423481&RS=15423481.

[15] Japan Pat. No. 5992079, https://www.j-platpat.inpit.go.jp/ c1800/PU/JP-2015-160560/ 136A10EEE6803981A8FFD8240056397BD525A99DB3B0E0316AD3E0FB70D07D88/10/ja.

[16] *Japan Pat No.6086977*, https://www.j-platpat.inpit.go.jp/c1800/PU/JP-2015-514699/ 65C07FE8C8E041826D20E8F5B2AE17AB1D6E28C241C5D90A8A5FE452DB9D2C36/10/ja.

[17] Kaneko, Tomoko; Yamamoto, Shuichiro; Tanaka, Hidehiko. "CC-Case as an Integrated Method of Security Analysis and Assurance over Life-cycle Process," *International Journal of Cyber-Security and Digital Forensics* (IJCSDF), Issue (Vol. 3, No. 1).

[18] Kaneko, Tomoko; Yamamoto, Shuichiro; Tanaka, Hidehiko. *CC-Case as an Integrated Method of Security Analysis and Assurance*, ICCC2014.

[19] Kaneko, Tomoko; Yamamoto, Shuichiro; Tanaka, Hidehiko. *CC-Case as an Efficient Method of Assurance Case for the Security Risk Management*, Promac, 2014.

[20] Chung, L; Nixon, BA; Yu, E; Mylopoulos, J. *Non-Functional Requirements in Software Engineering,* Kluwer Academic Publishers, 2000.

[21] Yu, E. "i*" *i*homepage*, (online), available from <http://www.cs.toronto.edu/km/istar/>.

[22] Dardenne, A; Lamsweerde, AV; Fickas, S. "Goal-directed Requirements Acquisition," *Science of Computer Programming*, 1993, Vol. 20, 1993, pp. 3-50.

[23] Yamamoto, S; et al. "Actor Relationship Analysis for the i* Framework," In: J. Filipe and J. Cordeiro. *International Conference on Enterprise Information Systems. LNBIP*, 24, 2009, p. 491-500.

[24] *The Tropos Methodology*, http://www.troposproject.eu/.

[25] Bresciani, Paolo; Perini, Anna; Giorgini, Paolo; Giunchiglia, Fausto; Mylopoulos, John. "Tropos: An Agent-Oriented Software Development Methodology," *Autonomous Agents and Multi-Agent Systems*, May 2004, Volume 8, Issue 3, pp. 203–236.

[26] *Ensuring Safety in Complex Systems*, https://www.ipa.go.jp/english/sec/complex_systems/stamp.html.

[27] Schneier, Bruce. "Attack Trees," *Dr. Dobb's Journal of Software Tools*, 24(12), (1999), 21–29.

[28] Sindre, Guttorm; Opdahl, L. Andreas. "Eliciting security requirements with misuse cases," *Requirements Engineering*, Vol.10, No. 1, pp. 34–44, (2005).

[29] Lipner, Steve; Howard, Michael. *The Trustworthy Computing Security Development Lifecycle*, https://msdn.microsoft.com/en-us/library/ms995349.aspx.

[30] Shostack, Adam. *Threat Modeling: Designing for Security*, Wiley., 2014.

[31] Wei, Jingxuan; Matsubara, Yutaka; Takada, Hiroaki. *HAZOP-Based Security Analysis for Embedded Systems: Case Study of Open Source Immobilizer Protocol Stack.*

[32] Tondel, Inger Anne; Jaatun, Martin Giljeand; Meland, Per Hakon. "Security Requirements for the Rest of Us: A Survey," *IEEE Software*, pp. 20-27, January/February, 2008.

[33] Liu, Lin; Yu, Eric; Mylopoulos, John. *Security and Privacy Requirements Analysis within a Social Setting*, RE2003.

[34] Taguchi, K; et al., "Curriculum Design and Methodologies for Security Requirements Analysis," *Progress in Informatics*, No. 5, pp. 19-34, 2008.

[35] Kaneko, Tomoko; Yamamoto, Shuichiro; Tanaka, Hidehiko. "A Spiral Review Method for Security Requirements," *ProMAC*, 2010, p. 1227-1238.

[36] Kaneko, Tomoko; Yamamoto, Shuichiro; Tanaka, Hidehiko. "Specification of Whole Steps for the Security Requirements Analysis Method (SARM) - From Requirement Analysis to Countermeasure Decision," *ProMAC*, 2011.

[37] Young, William; Leveson, G. Nancy. "Systems Thinking for Safety and Security," *Proceedings of the 29th Annual Computer Security Applications Conference (ACSAC 2013)*, 2013, pp. 1-8.

[38] Young, William; Porada, Reed. "System-Theoretic Process Analysis for Security (STPA-SEC): Cyber Security and STPA," *2017 STAMP Conference*.

[39] *ISO/IEC15026-2-2011*, Systems and Software engineering-Part2:Assurance case.

[40] *OMG, ARM*, http://www.omg.org/spec/ARM/1.0/Beta1/.

[41] *OMG, SAEM*, http://www.omg.org/spec/SAEM/1.0/Beta1/.

[42] Kelly, T; Weaver, R. "The Goal Structuring Notation – A Safety Argument Notation," *Proceedings of the Dependable Systems and Networks 2004 Workshop on Assurance Cases*, July 2004.

[43] Goodenough, J; Lipson, H; Weinstock, C. *Arguing Security - Creating Security Assurance Cases,* 2007. https:// buildsecurityin.us-cert.gov/bsi/articles/knowledge/assurance/643-BSI.html.

[44] Alexander, R; Hawkins, R; Kelly, T. *Security Assurance Cases: Motivation and the State of the Art,*, CESG/TR/2011.

[45] Kaneko, Tomoko; Yamamoto, Shuichiro; Tanaka, Hidehiko. "Proposal on Countermeasure Decision Method Using Assurance Case And Common Criteria," *ProMAC*, 2012.

[46] Kaneko, Tomoko; Yamamoto, Shuichiro; Tanaka, Hidehiko. "An Integrated Method of Security Analysis and Assurance using Common Criteria-based Assurance Case Using lifecycle-model," *ComSec*, 2014

[47] Yamamoto, Shuichiro; Kaneko, Tomoko; Tanaka, Hidehiko. "A Proposal on Security Case based on Common Criteria," *Asia ARES*, 2013.

[48] *Common Criteria for Information Technology Security Evaluation*, http://www.commoncriteriaportal.org/cc/.

[49] *NISC*, www.nisc.go.jp/conference/seisaku/dai15/pdf/15siryou02.pdf.

[50] NISC, *General Framework for Secure IoT Systems*, https:// www.nisc.go.jp/eng/pdf/iot_framework2016_eng.pdf.

[51] Information Promotion Agency (IPA), *IoT Safety/Security Design Tutorial*, https://www.ipa.go.jp/files/000053921.pdf.

[52] Information Promotion Agency (IPA), *IoT Safety/Security Development Guidelines* (Second Edition), Important Points to be understood by Software Developers toward the Smart-society https://www.ipa.go.jp/english/sec/reports/20160729-02.html.

[53] Information Promotion Agency (IPA), *Guidance for Practice Regarding IoT Safety/Security Development Guidelines* [IoT High Reliability Functions], https://www.ipa.go.jp/english/sec/reports/20171226.html.

[54] Kaneko, Tomoko; Takahashi, Yuji; Okubo, Takao; Sasaki, Ryoichi. "Threat analysis using STRIDE with STAMP/STPA," *The International Workshop on Evidence-based Security and Privacy in the Wild (APSEC2018Workshop)*.

[55] Friedberg, Ivo; McLaughlin, Kieran; Smith, Paul; Laverty, David; Sezer, Sakir. "STPA-SafeSec: Safety and security analysis for cyber-physical systems," *Journal of Information Security and Applications*, 2017, Volume 34, Part 2, pp. 183-196.

[56] Microsoft, Azure, *Internet of Things security architecture*, https://docs.microsoft.com/ja-jp/azure/iot-accelerators/iot-security-architecture.

[57] Soka Gakkai English Buddhist Dictionary Committee. (2002). "Ten factors of life," in *The Soka Gakkai Dictionary of Buddhism*. Tōkyō: Soka Gakkai. ISBN 978-4-412-01205-9. Archived from the original on 2016-02-26.

[58] Kaneko, Tomoko; Yoshida, Kazuki; Yoshioka, Nobukazu. "The modeling pattern of human and society for AI business," *Asian PLoP*, 2019.

[59] Kaneko, Tomoko; Yamamoto, Shuichiro; Tanaka, Hidehiko. "Efficient Use of Assurance Case against System Risk in the Project Management,"*ProMAC*, 2013.

[60] Kaneko, Tomoko; Yabe, Satoshi. "PS-Case –A Problem Solving Method Based on a set of Evidence," *Promac*, 2015.

BIOGRAPHICAL SKETCH

Tomoko Kaneko (PhD, Informatics)

Affiliation:
- NTTDATA Corporation
- Visiting Researcher of Institute of Information Security
- Researcher of National Institute of Informatics
- Researcher of Tokyo Denki University

Education:

3/2014 Doctor of Informatics, Institute of Information Security, Tokyo, Japan, Thesis title: "CC-Case as an Integrated Method of Security Analysis and Assurance".

3/2010 Master of Informatics, Institute of Information Security, Tokyo, Japan.

3/1992 Bachelor of Law, Soka University, Tokyo, Japan.

3/1988 Bachelor of Social Studies, Keio University, Tokyo, Japan.

Research and Professional Experience:

2017- present   Safety & Security Section Chairperson, Software Quality Management Research Group of the Union of Japanese Scientists and Engineers (JUSE)

2019- present   Project Researcher of QAML(Quality Assurance of Machine Learning) Project, Theme: Modelling and AI for Integration of Cyber and Physical World(JST Future Creation Project) https://qaml.jp/en/

2019- present   Assistant Secretary of Knowledge-based Software Engineering(KBSE) of The Institute of Electronics, Information and Communication Engineers(IEICE)

2019- present   Executive Committee, Secretary of IT Risk Studies, Japan Society of Security Management

2018- present   Part-time lecturer, IT Theory course, Tokyo University

2016- present   Part-time lecturer, Safety/Security Requirement course, Top SE of NII

2015   Part-time lecturer, Common criteria (ISO / IEC 15408) as Development method, Cysec of TDU

Professional Appointments:

1988 -present Deputy Manager, Technology and Innovation General Headquarters, NTT DATA Corporation Involved in the development and operation of many systems like Air to Ground Data Communication, Financial Futures Exchange, and Amusement

Equipment, or Engaged in Quality Assurance and System and Security Auditing.

4/2016- 3/2019 Information-Technology Promotion Agency (IPA), Japan, being sent to IPA from NTTDATA Corporation

Honors:

2018	Technical Encouragement Award, The Union of Japanese Scientists and Engineers (JUSE), Software Quality Management Research Group
2015	Excellence Award, 15th Anniversary Papers, NPO Japan Network Security Association (JNSA)
2015	Certified Auditor of Information Security (CAIS): Registration Number: B1511014410
2010	Fighting-spirit award, Institute of Information Security
2006	Excellence Award, "Telework Promotion Award" of Japan Telework Association
2013	ITIL Foundation Certificate
2013	COBIT Foundation Certificate
2013	CMMI for Development CMMI for Services supplement Certificate
2010	MEXT accredited program and certificate (Software Specialist)
2010	MEXT accredited program and certificate (Information Security Specialist)
2006	ISO9001 internal auditor and instructor qualification
1992	Class II Information Technology Engineer

Publications from the Last 3 Years:

Peer reviewed:

[1] Tomoko, Kaneko; Kazuki, Yoshida; Nobukazu, Yoshioka. "The modeling pattern of human and society for AI business," *Asian PLoP*, 2019, (in Japanese).

[2] Tomoko, Kaneko; Takahashi, Yuji; Takao, Okubo; Ryoichi, Sasaki. "Threat analysis using STRIDE with STAMP/STPA," The

International Workshop on Evidence-based Security and Privacy in the Wild (APSEC2018Workshop), (in English).

[3] Tomoko, Kaneko; Takahashi, Yuji; Takao, Okubo; Yoshimi, Teshigawara; Hidehiko, Tanaka. Evaluation Practice for the Effectiveness of CC-Case as an Integrated Method of Security Requirement Analysis and Assurance, *IPSJ Transactions Consumer Devices & Systems*, Volume 8, Issue 1, Page: 11-26, (WEB ONLY), 2018, (in Japanese).

[4] Tomoko, Kaneko. Visualization of Security Requirements by CC-Case_i for more Secure System Realization, *Journal of Japan Network Security Association (JNSA)*, Volume 30, No. 1, 2016, (in Japanese).

Non-reviewed:

[1] Tomoko, Kaneko; Takahashi, Yuji; et al., "Technology Integration Proposal for Security & Security Development-Integration of STAMP/STPA and Assurance Case," Technical Encouragement Award Paper, *The Union of Japanese Scientists and Engineers (JUSE)*, 2019, (in Japanese).

[2] Tomoko, Kaneko; Takuo, Hayakawa; Takahashi, Yuji; Takao, Okubo; Ryoichi, Sasaki. "Proposing enhancement of a threat analysis from the security perspective for STAMP/STPA as a safety analysis," *IPSJ SIG Notes* (Web), Volume: 2018, Issue: CSEC-80, Page: Vol. 2018-CSEC-80, No. 6, 1-8 (WEB ONLY), 2018, (in Japanese).

[3] Tomoko, Kaneko; Takahashi, Yuji; Takao, Okubo; Ryoichi, Sasaki. "Application of Threat Analysis (= STRIDE) to Safety Analysis Method STAMP/STPA," *Information Processing Society of Japan/ Information Processing Society of Japan Symposium Series (CD-ROM)*, Volume: 2017, Issue: 2, Page: ROMBUNNO.3C3-2, (in Japanese).

[4] Tomoko, Kaneko; Takahashi, Yuji; Takao, Okubo; Yoshimi, Teshigawara; Hidehiko, Tanaka. "*Proposal of an IoT Security Certification Method Using CC-Case,*" IPSJ Symposium Series (CD-

ROM). Volume: 2016. Issue: 2. pages: ROMBUNNO.1E3-2, 2016, (in Japanese).

[5] Tomoko, Kaneko; Takahashi, Yuji; Takao, Okubo; Yoshimi, Teshigawara; Hidehiko, Tanaka. "Visualization of IoT Security Requirements Using CC-Case," *IPSJ SIG Notes* (Web), Volume: 2016, Issue: MBL-80, Page: Vol. 2016-MBL-80, No. 5, 1-8. (WEB ONLY), 2016, (in Japanese).

Book Chapters:

[1] *STAMP Guidebook: Safety Analysis by System Thinking, Information-Technology Promotion Agency (IPA)*, 2019, (in Japanese).

[2] *STAMP/STPA for beginners for beginners (Utilization version)-Future Safety by System Thinking*, IPA, 2018, (in Japanese).

[3] *STAMP/STPA for beginners (Practice version)-New Safety Analysis Method based on System Thinking*, IPA, 2017, (in Japanese).

[4] *Guidance for Practice Regarding "IoT Safety/Security Development Guidelines" [IoT High Reliability Functions]*, Information-Technology Promotion Agency (IPA), 2017, (in English) https://www.ipa.go.jp/files/000063228.pdf.

Articles in Nonarchival Magazines or Journals:

[1] Tomoko, Kaneko; kiyoshi, Nakazawa. "Current Status of Safety Analysis Method" STAMP, *JETRO News*, May 2018, (in Japanese).

[2] Tomoko, Kaneko. "Security by Design and Assurance Case," *SEC Journal*, 47, P28-33, 2016, (in Japanese).

Invited Presentations:

[1] *JR East Technical Seminar Invited Lecture*, "Quality Assurance of Security by Assurance Case and Common Criteria (ISO/IEC 15408)," 5/2018.

[2] *Invited Presentation of CEATEC Japan 2017*, "Security by Design for the realization of a Safe and Secure IoT Society," 10/2017.

[3] Invited Lecture of Threat Analysis Study Group "Applying Security for STPA," 8/2018 *Invited Presentation of NSPICE Conference 2017*, "Safety Security Development and Assurance Case," 10/2017.

In: A Closer Look at Safety and Security
Editor: Jeff M. Holder

ISBN: 978-1-53618-176-0
© 2020 Nova Science Publishers, Inc.

Chapter 2

SAFETY AND SECURITY CHALLENGES PREVENTING PARENTS FROM ENROLLING CHILDREN IN EARLY YEARS SCHOOL IN DIFFICULT CIRCUMSTANCES

Nyakwara Begi[*], *PhD and Yattani D. Buna*[†], *PhD*
Department of Early Childhood and Special Needs Education,
Kenyatta University, Nairobi, Kenya

ABSTRACT

Early Years Education (EYE) is a basic right for every child as enshrined in the United Nations Convention on the rights of children. The African Charter on Rights and Welfare of Children, Kenya's Education Act 2013 and Children's Act 2012 emphasize that children's environment should be free from physical and socio-emotional stress that hinders them from accessing EYE. However, research has shown that there are many children in difficult circumstances not accessing EYE owing to safety and security concerns in the environment. In view of the importance of EYE, there was need to carry out a study to establish the safety and security

---

[*] Corresponding Author's Email: begi.nyakwara@ku.ac.ke.
[†] Corresponding Author's Email: buna.yattani@ku.ac.ke.

challenges preventing parents from enrolling their young children in schools with the intention of recommending strategies to address the challenges. This study was guided by the ecological systems theory which explains how factors within the environment influence children's development and education. Descriptive survey research design was used; while data was collected using interviews and focus group discussions. The study was done in Marsabit County which is one of the counties in Arid and Semi-arid Lands (ASAL) in Kenya. The target population was school going children and their parents in households in the county. Purposive and random sampling techniques were used to select parents, teachers, area education officers and chiefs. The research instruments used for collecting data were interview schedules and focus group discussions which were analyzed using qualitative and quantitative techniques. Results from data analysis show that several safety and security challenges were preventing parents from enrolling their children in schools. The challenges included: Long distance from home to school; harsh weather; rugged terrain; salty water; and ethnic conflicts. Some of the recommended strategies to address the challenges include establishment of mobile schools near villages; flexible school routine; avoiding rugged terrain when selecting sites for establishment of villages; boiling and filtering water for drinking; and Organising frequent peace and reconciliation meetings.

Keywords: safety, security challenges, preventing, parents, access to early years education, difficult circumstances

1. INTRODUCTION

Early Years education is a basic right for every child as enshrined in the United Nations Convention on the Rights of the Child. The convention underscores education as the right which all governments must strive to provide to all children. The World Education Forum on Education For All (EFA) goals was committed to provision of a comprehensive early childhood education and care for the vulnerable and disadvantaged children (UNESCO, 2007). This is because children's access to school during early years exposes them to early screening and intervention which reduce prevalence rate and hence saving resources that would have been used on treatment services (Organisation for Economic Co-operation and

Development [OECD], 2016). Research has shown that access to early years education for disadvantaged children reduces school dropout, crimes, conflicts and poverty (Ramey, 2000). It is therefore apparent that denying children in difficult circumstances access to early years education make them more deprived and not to reach their full potential (Duncan et al., 2007). It is also clear from research that investment in early years education brings greater returns compared to investment in later years of education (Heckman & Masterov, 2004). This means that one of the strategies that can be used to reduce poverty in marginalized areas is by ensuring that all children have access to early years education.

Globally, children's access to education in early years is an issue of concern in many countries. According to Australian Early Development Census Report (2016), children from disadvantaged backgrounds were more than twice likely to be developmentally vulnerable at the start of school and the enrolment rate of the aboriginals children lagged behind those children of non-aboriginals in South Wales in Australia (State of Early Learning in Australia, 2017). A study done in the USA by McDonald (2005) revealed that many children did not live within walking distance to school therefore, prompting the use of bus services leading to many cases of child obesity. In a cross-sectional study conducted in the United Kingdom by Panter, Jones, Sluijs and Griffin (2011) to examine the relationship between active commuting behaviour, levels of physical activities and distance to school among 9-10 year old children revealed that long distance to school in dry areas made children feel exhausted and hindered their access to school.

In Asia studies have been conducted on various factors influencing children's enrolment in school. Ranabhat (2014) conducted a study on determinants of access, participation and learning outcomes at primary school level in Nepal. The findings showed inequitable enrolments and as well as disparities in access and participation among ethnic minority groups and that there were both spatial and social disparities in access making EFA achievement far from success by 2015. In Indonesia, Korkeala (2011) found that there was reduced number of children continuing from primary to secondary school during the onset of monsoon season and delayed entry to school for young children and as well as increased child labour. A study conducted in Afghanistan by Burde and Linden (2012) on the effect of Village Based Schools observed that distances between villages were greater and travelling between them was dangerous for young children and girls. It is therefore clear from the studies that long distance from home to school was a challenge affecting children's access to school.

Studies conducted in Africa related to children's enrolment in early years education revealed interesting results. In Nigeria, Duze (2010) carried out a study on the average distance children travelled to schools and its effects on school attendance. Results showed that students travelled up to five kilometres and that majority of children travelled more than the recommended one kilometre distance to school which affected school attendance. Vuri (2008) when assessing the effect of long distance from home to school on children's work performance and school attendance in Ghana and Guatemala found that the distance from home to school had negative effect on children's work performance at home and school. This was because the longer distance children travelled to school the more difficult it was for them to reconcile work and school attendance. A study carried out by Bongo (2015) in Zimbabwe on the potential effects of disasters on children's access to quality education observed that floods caused loss of learning hours, absenteeism of children from school, outbreak of diseases and low coverage of the syllabus culminating in poor academic performance. The study further reported food insecurity and dropping out of school which lead to early marriages for girls. It is therefore clear that long distance between home and school, insecurity and safety of children were some the factors impeding children's access to school results which are similar to the findings from studies conducted in Asia.

In Kenya, the constitution and Education Act 2013, expresses the right of children to participate in education (Republic of Kenya [ROK], 2010, 2013). In spite of these legislations, access to early years education for the nomadic pastoralist children continues to trail behind other parts of the country (Abdi, 2010). Literature reviewed had shown that in Marsabit

County children's access to early years education was low (Uwezo 2011). Ruto, Ongwenyi and Mugo (2010) in a report titled Educational Marginalization of the Arid North; observed that marginal arid and semi-arid districts of Northern Kenya had low participation rates. A study carried out by Mulinge (2012) on factors affecting urban refugee children's access to education in public primary schools in Nairobi County found that factors such as language barriers, unsafe living conditions and hostile social environment were factors that hindered children's access to education. Gitau (2013) carried out a study to investigate the prevalence of drought and its impact on the learning of pupils in Laikipia West County. The study revealed that drought period was recurrent in the area and often led to children's absenteeism, poor performance, truancy and school dropout.

In view of the importance of early years education, the current study sought to explore the safety and security issues preventing parents from enrolling their children in early years school in ASAL. The study also sought to explore the strategies that can be put in place to address the safety and security concerns to encourage parents to enroll their children in early years schools.

2. Problem Statement

Literature reviewed show that early years education is a basic right for every child including those living in difficulty circumstances. Children's access to early years education gives them a strong start, enhances smooth transition to primary school, reduces school dropout, community conflicts and poverty. Denying children in difficult circumstances access to early years education make them more disadvantaged and hinders them from reaching their full potential.

It was also clear from the literature reviewed that pastoralists' communities continued to register low enrollment of children in early years school. Research studies revealed that the ASAL areas had unique challenges that impacted negatively on children's access to education.

Thus, there was need to explore the safety and security issues preventing parents from enrolling their children in early years school.

3. PURPOSE OF THE STUDY

The purpose of this study was to establish the safety and security issues preventing parents from enrolling their children in early years school as well as to explore the strategies that can be put in place to address the issues in order to enhance children's enrollment in early years school.

4. OBJECTIVES OF THE STUDY

1. To establish the safety and security challenges preventing parents from enrolling their children in early years school in difficult circumstances.
2. To determine the strategies that can be put in place to address the safety and security challenges to enhance children's enrollment in early years school.

5. THEORETICAL FRAMEWORK

This study was guided by the ecological systems theory by Urie Bronfenbrenner (1979) based on five concentric systems or levels. These levels are Microsystem, Mesosystem, Exosystem, Macrosystem and Chronosystem. To him, child development is a process that takes place within and across different settings. He observed that the interactions within each level and across the settings do have a considerable influence on the child's development and education. He used this theory to explain the growth, development and education of children. Bronfenbrenne state that each of the factors within the different levels have an impact on each other and on the child. Therefore, there are many factors that influence children's access to early childhood education which may include parental attitudes, cultural beliefs, family income or occupation and government policies and ideologies.

Macrosystem refers to the environment in which the child lives. The environment comprises the family, friends, peer group, neighbourhood and the school. These groups interact with the child and influence the child's development. It is from this immediate setting that the children develop attachment, friendship and even identify their own roles in life. Each of these settings has a role to play as far as enhancing the child's access to school is concerned. The school is an important institution where acquisition of knowledge takes place. Teachers on their part have to be good role models and also make schools favourable setting to win parental approval to enroll their children in the school.

Mesosystem involves the interactions between the factors within the immediate environment. According to Bronfenbrenner, the interaction between the people at home and school tends to influence one another. Each setting has a role to play for the best interest of the child. The family sends children to school and teachers impart skills and knowledge. They are all interdependent. The parents, therefore, have the role of interacting with teachers, children's friends and other parents, school committees in enhancing both the home and school environment. It is the role of the parents to interact with stakeholders within their environment and to enroll their children in school and to maintain good rapport with teachers.

Exosystem is a setting which is beyond the child's immediate environment but has considerable influence on the child. It consists of parents' employer, local government, the community, social welfare and mass media. The difficulties that the parents experience at their workplace which can lead to loss of a job will have serious impact on the child in terms of low income for the family which may translate into lack of basic necessities and fees for children. Therefore, it is important to observe that the Exosystem environment such as parents' work is taken care of and therefore government decisions can influence children's access to early childhood education.

Macrosystems comprise the society's socio-cultural practices, attitudes and ideologies, which have great influence on children while the children have no control over them. The socio-cultural values of the community may put greater demands on the child. For instance; early marriage, Female Genital Mutilation (FGM) and cultural ritual observance which are from the family and the community may impact on children's access to ECE programmes.

Chronosystem refers to experiences a person gathers through life. These experiences carry with them events that offer lessons from which individuals learn to understand their environment develop independence and contribute positively to their own development. It is also important to point out that it is not only the child going through change but the environment of the child also changes. Cultural values are overtaken by Information Technology age where the world is becoming a global village. Parents and children will embrace this change with time and the government will be expected to develop policies that will enhance children's access to early childhood education in Turbi Division, Marsabit North Sub-County.

Bronfenbrenner (1992) asserts that when all the different factors interact in harmony with the child, they will enhance the child's development and education. The parents, stakeholders and peers interact to create awareness on the need for enrolling children in school. Other factors such as parental attitudes, income levels, socio-cultural practices, government policies and technological changes can also help policy-makers to invent new ways of making education accessible to children especially for those hard-to-reach children of nomadic pastoralists, for instance, digital learning.

This theory is relevant to this study as it explains the importance of the different factors within and outside the child's environment which can either enhance or hinder his/her access to education. This theory helps to explain how the parents (their occupation and income, educational level) other stakeholders, government policies, geographical factors and cultural values interact to influence children's access to early childhood education.

6. Research Methodology

It has been described under the following sub-sections:

6.1. Research Design

This study adopted a descriptive survey research design. According to Robson (2004), descriptive study involves use of extensive previous knowledge of a situation so as to give an accurate profile of a person, events or situation. This design was appropriate for the study since the researchers wanted to establish the safety and security issues preventing parents from enrolling their children in early years school as well explore the strategies that can be put in place to address the issues in order to enhance children's access to school.

6.2. Location of the Study

The study was done in Marsabit County which is located in northern Kenya on the eastern shore of Lake Turkana. It is the largest county in Kenya covering 70, 961 square kilometres. It borders four counties; Wajir to the east, Turkana and Samburu to the west, Isiolo to the south and Ethiopia to the north. The county lags behind other counties in terms of education and other services. Studies have also revealed that more than half of three to five-year old children were not attending school (Uwezo, 2011).

Regarding the climatic conditions, the area is hot with an average of 36 degrees Celsius during the hot months and 25 degrees Celsius during the cold months. The average precipitation received is 254mm making it one of the driest counties in the Kenya. Most of the rainfall is received in March and November however on average six days of the month the area experiences foggy conditions due to its proximity to the lake and other geographic features such as Marsabit Mountain. This makes it one of

Kenya's driest counties. Most of the rainfall is received in April and November. These temperatures influence decision of parents to enrol young children in school especially where the children have to walk long distances under hot sun to and from school.

According to the Commission for Revenue Allocation (CRA) the constituency poverty ranking in 2009 was reported to have 75% of the population living below the poverty line. This area was purposively sampled for this study because it was predominantly inhabited by the nomadic pastoralist communities and the area often experiences prolonged drought leading to livestock deaths which may aggravate the levels of poverty and which may likely affect children's access to early childhood education.

6.3. Target Population

The target population was school age going children and parents in 156 households in five villages, head teachers, teachers, area chiefs and the Area Education Officers in Turbi Division.

6.4. Sampling Techniques and Sample Size

Purposive sampling was used to select Marsabit County and Turbi Division. This was because the method allowed the researchers to select the sample which had the required information and was to serve the purpose of the study. When the population is relatively huge, Sutter (2011) suggests that the use of a 30% to 60% sample of the total population is appropriate in educational studies. Hence, purposive sampling was used to select 50% of the villages to be involved in the study. The households were numbered and then simple random sampling was used to select 50% households from the identified villages. All head teachers, teachers, area chiefs and education officers in the sampled area were included in the study.

6.5. Sample Size

It consisted of 78 parents from the sampled households, four headteachers from the four schools in the sample area, twelve teachers, two chiefs and two Area Education Officers (AEO). This gave a total sample of 98 respondents.

6.6. Research Instruments

The instruments used to collect data were interview schedules and focus group discussions. The instruments were used because the majority of the parents and the chiefs were illiterate or had very little education. Interview schedules were also used to collect data from headteachers, teachers and AEO to corroborate findings from parents. The focus group comprised of parents, head teachers, teachers, the chiefs, and Area Education Officers. Validity and reliability of the instruments were done during pilot study.

6.7. Data Collection

Data was collected in four stages and analyzed using qualitative and quantitative methods. The results were presented using frequencies, percentages and texts. FGD was conducted after all the interview sessions were over in order to generate information to corroborate responses obtained through the interviews.

6.8. Logistical and Ethical Considerations

Authority to conduct the study was obtained from different relevant authorities before undertaking the study. The researchers also had obtained consent from the participants and were assured of their right to withdraw from the study at any time in the research process. They were also informed that the information gathered was to be kept confidential and used for the research purpose. The identity of the respondents was also concealed as names were not used.

7. Results and Discussions

It has been described under the following sub-sections:

7.1. Safety and Security Issues Preventing Parents from Enrolling Children in Early Years Education

The first objective of the study was to establish the safety and security issues preventing parents from enrolling children in early years school in difficulty circumstances. To establish the issues, parents, headteachers, teachers, chiefs and area education officers were interviewed. Similarly, a focus group discussion was conducted to corroborate the findings. Results from data analysis revealed several issues that prevented parents from enrolling their children in early years school. The main issues have been discussed under the following sub-headings:

7.1.1. Long Distance to School

Long distance from home to school was cited by many respondents as a safety and security issue preventing parents from enrolling their children in school. A parent from a remote mobile settlement had this to say:

> We live far away from the town with schools and our livestock have huge labour demands where we rely on our children to help in taking care of some of the livestock. Big children will look after and care for camels and goats while younger children will look after the young camels and goats. Girls may remain at home to care for younger siblings. I have enrolled my third born son in school when he was old enough to be in a boarding school but the distance, our nature of mobility and the non-

existent of early childhood centres could not allow us to enroll even those whom we may like to enroll in school (Parent 2, Village Z).

Headteachers, teachers, chiefs and area education officers interviewed had also expressed that long distance from home to school prevented parents from enrolling children in school. One headteacher had described the situation as follows:

> The distances between settlements and schools was bound to vary as the pastoralist movements were dependent on weather patterns and governed by the availability of basic needs, such as water and pasture for their livestock. There were times these people settle closer to schools but they could not enroll their children in school due to livestock or other domestic considerations (Headteacher, School 3).

The Focus Group Discussion also revealed that the mobility of pastoralists and frequency with which pastoralist settlements relocate to new sites in pursuit of the needs of their livestock, always keeps them at great distance from schools. The participants unanimously concurred on the notion that long distance from school prevented parents from enrolling their children in school particularly for young children who were not able to walk long distances or even stay in boarding school.

A parent living in a close proximity to school made the following comment on why she did not enroll her children in school:

> Even though I live about 2 kilometres from the existing pre-school, I did not enroll my children because between the school and the village there is a road corridor leading to a watering point along which there is heavy livestock traffic going to watering point. Indeed, enrolling the child in the pre-school will require someone to take the children to school and to pick the child after school and the family members are so committed elsewhere to take the child to and from school daily. So we decided to retain the child home until such a time when he has come of age to go to school alone (Parent 19, Village V).

The implication that could be drawn from the above results is that there were other factors other than distance that hindered children's enrolment in school besides long distance between home and school. This means that the school can be situated near a settlement and the parents may not still enroll their children in school due to their own other considerations.

7.1.2. Harsh Weather

The harsh weather was the other main safety and security issue preventing parents from enrolling their children in early years school in difficulty circumstance. It was reported that high day temperatures and extreme wind filled with dust particles make parents delay or not to enroll young children in school. This view was expressed by the parents living 3-6 kilometres from school. They hold the view that the intense heat makes children thirsty, sweat profusely and dust make children fall sick and hence reasons for which they were not enrolling their children in early childhood education even for those not living far from the school.

One parent, a mother in a village situated between three to five kilometers from an early childhood centre had described the situation as follows:

> Here in this area the mid-day sun is so hot and children get thirsty and sweat a great deal. Walking home from school is a hazardous undertaking for young children. That is the reason why we find it hard to enroll children in school to walk in the extreme heat of the sun. (Parent 6, Village W).

Similarly, another parent residing in the same village had also reported as follows:

> Some days the weather becomes so windy and filled with dust. So walking long distances in the dust can make them get cough and eye infections so that is the reason why I did not enroll my children in school until they are able to protect themselves from the dust (Parent 10, Village W).

The interview with headteachers, teachers and chiefs also concurred with parents that unfavourable weather elements such as high temperatures made some parents not to enroll their children in early childhood centres. Similarly, Focus Group Discussions also agreed with the above sentiments but expressed that parents preferred enrolling older children in pre-school centres as they believed that younger children were vulnerable.

7.1.3. Rugged Terrain

This is the other safety and security issue preventing parents from enrolling their children in early years school in difficulty circumstance. Some parents had cited fear of rugged rocky terrain and the perceived challenge it poses to their young children as one of the main reason why they did not enroll their children in school. When parents were asked why they did not make roads to their village, they cited some challenges in creation of roads every time they moved to a new site. One male parent during the interview said:

> One reason why I could not enroll my pre-school age child in school besides the distance is the fact that where we are now, the area is dominated by rocky terrain where both children and adults are easily prone to injuries; where they could easily sprain or even fracture legs and arms. Also I don't see the need to make roads every time we settle in a new area as it is a costly affair. Such situations make me not to send my young children to school as at now but in future, God willing. (Parent 15, Village V).

The other respondents: Headteachers, teachers, chiefs, and area education officers also identified rugged topography in some areas as a factor which was an issue preventing parents from enrolling their young children in schools in difficulty circumstance. They said that the rough landscape posed walking difficulties to young children and fear of accidents make parents choose not to take their young children to school. One pre-school teacher made the following statement:

> The terrain in some areas are rocky grounds which posses difficulty for people to walk on such terrain. Children and adults alike are prone to accidents on such rough topography and out of fear of such accidents parents tend to hold back their children until when they are of age to walk with ease on such terrain.(Parent 5, Village X).

7.1.4. Ethnic Conflicts

Northern Kenya is an area predominantly inhabited by nomadic pastoralist communities. These communities differ in cultural practices and sometimes even language. However, the commonality they share across all of them is that livestock forms the source of livelihood. Each and every one of them committed to selflessly safeguard the welfare of their source of livelihood; that is camels, cattle, goats, sheep and donkeys. Their life routine is centered on cyclic pattern with overriding priority of provision of pasture and water for their livestock. In their search for pasture and water for their livestock conflict often arise among these communities. Such conflicts sometimes culminate to full scale war leading to displacements of people and families. During such conflicts fleeing families carry with them their children leading to closure of schools. On the worst scenario, women and children are targets of the groups and hence making parents not to enroll their children in schools.

Ethnic rivalry in Northern Kenya has also been said to have political undertones. It is a common trend for conflicts to occur either before elections or after elections. Before elections it's aimed at causing fear and tension among the opponents in order to disrupt voter turnout and to influence the outcome to their advantage. Other time conflicts arise after election in order to give the winner difficult time and disrupt them from carrying out any meaningful development. These conflicts whether instigated by politicians or scramble for scarcity of pasture and water often leads to closure of schools for a long time and hence making parents not to enroll or withdraw their children from school.

7.1.5. Saline Water

Similarly, in some Arid and Semi-arid areas, the common sources of water are the shallow wells and water pans. Some of these sources are often salty and hence not favourable for domestic use. It was also reported that often children complain of abdominal pains and burning sensations when passing urine. They also suffer bone malformation, coloured teeth and deformed leg/limbs. Parents often avoid taking their children to schools in centres with high concentration of saline water and prefer schools in areas with no salty water.

The findings from this study are similar to those of other studies. For instance, Sanou and Aikman's (2005) who did a study on pastoralist schools in Mali observed that the region occupied by Touareg was a semi-desert area which often suffered severe drought. Similar to the current study findings, children of Mali pastoralists also have to walk long distances (4-8 Kilometres) to and from school daily. The long distances to school may tend to militate against children's access to school. The study findings support the findings of the current study where distance and high day temperatures hinder children's access to early years school. Fernandez-Gimenez, Batkhishig and Batbuyan (2012) carried out a study on children's vulnerability and capacity building adaptation in Mongolia. The study observed that extreme weather patterns and environmental hazards such as drought, snow, dusty storms and encroaching desertification were some of the global climate changes affecting children and recommended the need to cushion children against such vulnerability, which if not checked deny children access to school.

The study findings were further supported by other studies conducted in Afghanistan by Burge and Linden (2008) who observed that distances between villages were greater and travelling between them was dangerous for young children and girls. Duze (2010) found that in Nigeria, learners travelled long distances to school which affected their school attendance. Similarly, Bongo (2015) found that in Zimbabwe long distance between home and school, insecurity and safety of children affected children's access to school. Mulinge (2012) and Gitau (2013) in Kenya found that unsafe living conditions and hostile environment hindered children's access to education. Gakuria (2013) observed that most tribal conflicts in Kenya were caused by lack of adequate resources and change of climate which make communities to fight for the limited available resources. It is therefore clear from the finding that long distance from home to school, harsh weather, rugged terrain, salty water, and ethnic conflicts were some of the safety and security challenges preventing parents from enrolling their children in schools.

7.2. Strategies to Address Safety and Security Issues Preventing Parents to Enroll Children during Early Years of Education

The researchers were also interested to explore the strategies that can be used to address the safety and security issues preventing parents from enrolling young children in school. To realize this objective, parents and the other participants were asked to suggest solutions the issues. Some of the strategies have been described in the following sub-sections.

7.2.1. Establishment of More Mobile Schools

From the interview with parents, it was noted that the number one strategy suggested by parents to improve children's enrollment in school was establishment of mobile schools. As pointed out earlier, some settlements were far away from the existing schools and there were no schools to serve children. In this regard, availing a pastoralist friendly school suited to the way of life of these people would be a worthy undertaking. One parent interviewed said:

> Our settlements are situated far away from schools situated in towns. We are on the move every season depending on the availability of pasture for our livestock. For this reason, there are no schools for our children near our settlement (Parent 3, Village X).

The interview with parents also revealed that there were some villages which used to have mobile schools and due to lack of teachers, they closed down. The parents had also reported that even though they were of low quality, the existing mobile schools and those that existed previously offered opportunity to their children to acquire early childhood education. One parent from a nomadic village with a mobile school remarked that:

> The mobile schools were on and off, the teachers were not paid and there was no coordination or help from outside as at now. But these mobile schools have assisted many of our children to access early childhood education. The first group of students are now in form one. We are grateful to the Catholic missionaries, UNICEF and our professionals who have managed the school even though they are far away in towns like Nairobi. We needed more help towards strengthening our mobile

schools because our teachers left; only one has remained (Parent 1, Village Y).

From the above views of parents, mobile schools have been quite handy because of their flexibility and proximity and the magnitude of work they have achieved since 2006 when they were established.

Sanou and Alkman (2005) who did a study to evaluate a Non-Governmental programme which aimed to improve enrolment of girls expressed that mobile schools were expensive to maintain which the participants of this interview responses seemed to disagree. Information gathered from interviews from Focus Group Discussion and observations revealed that mobile school structures were made from locally available materials and they resembled the cultural huts of the community and its construction do not involve much expenses. McMillan, (2007) carried out a study on the provision of education to the travellers' children in the United Kingdom expressed that the government in their endeavor to provide gypsy travellers' children with education, developed buses fitted with play boxes which contained complete package for all activity areas. The parents were inducted on the use of play boxes supported by photographic booklets for use by parents with low education when helping children. As pointed out elsewhere in this research, lack of teachers hampered children's access to early childhood education, and the use and availability of packaged teaching learning resources similar to those of the above findings will likely improve parental participation and enrolment of children in school.

According to Sengupta and Ghosh, (2013) in Indian cities like Mumbai and Delhi there were mobile bus classroom developed through the efforts of NGOs working in the area in conjunction with the Indian government. The schools on the wheels travelled to the children in the settlements during hours that suit their lifestyle and the nature of the occupation they were engaged in. In these model schools, classes took place during the normal hours of the day except in places where children were not available during the day. In such cases, classes occurred in the evening when children came home from fishing. Whereas the above study sought to develop appropriate school model compatible with the children of fishing community who were not available during the normal school hours, the current study sought to develop the appropriate movable school model suitable for children of the nomadic pastoralist who could not have access to school.

During the focus group discussion, the participants when brainstorming on the best strategies for enhancing children's access to school explored many options. One area education official had this to say:

> Because our people are on move every 3-4 times a year, it would be wise to get a movable classroom for children. it would be appropriate these classrooms can have wheel's, so that they are easily relocated to any site the nomadic satellite village move to such movable classes can be provided to all the pastoralist settlements to act as early childhood classrooms for the children of nomadic pastoralists (AEO 1, FGD).

The FGD also explored on how best this intervention programmes could be implemented for success. One teacher stated as follows:

> Documentation of the number of pastoralist households, their school age children and their movement patterns. The mobile schools should be coordinated from one central point. The time when any of nomadic pastoralist satellite village wish to migrate to a new site, they should send information early enough to the central point for the arrangement to be made in order to organize their movement. The classroom structure could be a "trailer with wheels' having adequate ventilation and spacious to accommodate 20-30 children that can be pulled to a new site using a truck (lorry). The organization coordinating should make prior arrangements to see to it that the classroom structures are relocated to a new site. To facilitate this movement the coordinating body should have a lorry to enable the movement of the classroom when need arise (Teacher, School 3).

The FGD expressed that for all the above intervention measures to be put in place, there should be support from key stakeholders including the government, NGO organizations and other donor support to ensure the success of the provision if the model schools for pastoralist children.

7.2.2. Government Support

The information gathered from parents showed that 48 (64.9%) of the parents had suggested government assistance through fees, bursaries and grants to schools as a major strategy for enhancing children's access to school. One parent from a remote settlement over 25 Kilometers from existing school, when asked in what way they expected the government to help them, responded in the following words:

> We live far away from urban centres with schools and grazing fields are distant from the towns. We needed government and other stakeholders to help educate our children through extension of fee bursaries and grants to our Early Childhood Education centres. (Parent 10, Village X).

Similarly, interviews with headteachers, teachers, chiefs and Area Education Officers had suggested inclusion of pre-primary school education into FPE policy so as to enhance children's access to pre-primary school for children who could not access early years education due to lack of fees. In support of the above point, Focus Group Discussion expressed that the government support should include all the necessities required in the school. One respondent explained as follows:

> Government support is very critical in terms of fees support, payment of staff salary and provision of teaching and learning resources. The Government both at national and county levels should be concerned with community education and enforcement of policies with regard to pre-primary education. (Parent 16, Village W).

7.2.3. Recruitment of More Teachers

Parents also suggested employment of early childhood teachers. As earlier revealed, there were few pre-schools in the sampled area of study and the construction of more pre-schools requires corresponding increase in the number of pre-school teachers. It is a common knowledge that schools without teachers will cease to operate and therefore, the need to employ more teachers. One parent from a mobile settlement which used to have a mobile school responded that:

> One time, we used to have a mobile early childhood centre in our village through the assistance of well-wishers. This helped many of our children to attend pre-school and later primary school. The teacher used to be from another distant town and after two years the teacher left the school citing family problems and little salary. The other challenge was that parents were not prompt in paying teacher's salary. We appeal to government and other stakeholders for help. (Parent 7, Village Z).

The headteachers, teachers, chiefs and the area education officers also expressed that early childhood centres will close their doors when teachers leave work due to low morale occasioned by poor pay. The Focus Group Discussions also indicated the need for the government and other stakeholders to employ more teachers with reasonable pay packages to retain them.

7.2.4. Collaboration with Stakeholders

Stakeholders were said to be key pillars in the establishment of schools, resource support and in monitoring and evaluation of learning in schools. Notably, the faith-based Organizations are the main backbone for support of education in the county. The Focus Group Discussion had expressed that stakeholder collaboration was an important strategy for expanding children's access to Early Childhood Education. A headteacher of one of the existing schools asserted that:

> Besides the County and National government support, help is also required from other stakeholders such as religious organizations, Non-Governmental organizations and the private sector. This should be done in a well-coordinated manner to avoid conflict of interest and duplication of roles (Headteacher 1, School 1).

The parents interviewed had expressed that the previously initiated mobile schools for the community collapsed because of lack of continued monitoring and assistance. One parent said:

> We have witnessed previous efforts through the churches and other well-wishers which established mobile schools for a few of our nomadic settlements. This did not last long because there was no support from other stakeholders. There was lack of follow-up and support structures in

place in terms of overseeing operations and as well as enrollment of children in school and teaching and learning regularly in the mobile schools (Parent 1, Village Y).

Similar views were expressed by the Focus Group Discussions. They suggested that proper planning and involvement of nomadic settlements to ensure future sustainability of mobile schools to enable access for children in a familiar environment within their settlements.

A study in Mali by Sanou and Aikman (2005) examined a Non-Governmental funded programme to improve gender equality in education through the work of "female community mobilizers" who were to support girls' access to education and to foster their participation through complementary developments designed to make the curriculum more gender equitable. The female community mobilizers were appointed to each school to work with parents and to enlighten them on the importance of education of their children and to monitor girls' attendance and ensure safe and girl friendly school environments. Such a model can be replicated among our pastoral nomadic population to enhance children access to quality early childhood education services with a follow-up of instance cases where children were not enrolled or were dropping out.

7.2.5. Integration of Indigenous Knowledge in Teaching and Learning

The interview with parents, headteachers, teachers and chiefs and the FGD participants suggested incorporation of aspects of local cultural values such as stories, play, riddles, tongue twisters, games, proverbs and as well allowing children's participation in cultural ceremonies. They argued that this would make parents who were skeptical about formal education gain some level of confidence and positive attitudes to enroll their children in early years school. One educated parent in the Focus Group Discussion expressed in his own words as follows:

> Teaching of some socio-cultural values are important because it will make children fully in touch with all aspects of their community's language, stories, children's play and the virtues of respect and socialization processes. This will make the parents gain confidence in school curriculum and feel that it has same objectives as their indigenous

education and therefore may attract those parents who were suspicious about formal education to embrace it and enroll their children in school (Parent 5, Village V).

The above revelations imply that incorporation of indigenous practices and values into the formal curriculum for children of this community could enhance their children's access to early childhood education. Studies done elsewhere have pointed out similar findings. Evans and Myers (1994) cited in Pence and Shafer (2006) expressed that incorporation of indigenous knowledge into school programmes would improve child care practices and as well as promote respect and appreciation for cultural values. Other studies have also observed that the use of child's mother tongue as a medium of instruction during early childhood period increases children's attendance in school and makes teaching-learning interesting (Begi, 2014).

A study by Andrew and Orodho (2014) in Kibera informal settlement had recommended increased government collaborative initiatives with different stakeholders through mobilization of physical, teaching and learning resources in the slums to enhance classroom instruction for children. The study further advocated for intensification of adult education programme to build the capacity of parents to value education and to assist their children with schoolwork at home. The study also recommended the initiation of Income Generating Activities (IGAs) to boost their economic stability and to support their children's education. The mobilization of resources for Nomadic Pastoralist parents was found to be important for boosting their children's access to school.

Bonds (2012) did an evaluation of the impact of school feeding programme on primary school enrolment in India. The results showed that the feeding programme leads to increased enrolment rates especially for children from low socio-economic status. Such programmes have been implemented in ASAL areas although some remote pre-schools and early childhood centres were not receiving such interventions from the data obtained from parents and Focus Group Discussions. Although not covered under this study school feeding programme has been widely believed to be beneficial to those children from poor vulnerable populations and those from geographically disadvantaged populations such as those of nomadic pastoralists in helping their children access early childhood education.

The current study findings were also supported by those reported by Miako (2012) who recommended government subsidies to help children of extreme poverty to access education in Kenya. The study also recommended for the employment of more teachers to cater for the increasing student's population and to provide more funds for the schools. A study by Resa (2001) on factors affecting enrolment and retention of children in primary school recommended increased budgetary allocation to enhance participation rates for children. Similarly, Migwi (2010) study on improvement of early childhood centres recommended Community Support Grant (CSG) and employment of teachers as some of the strategies for improving children's access to early childhood education. The study findings suggest that the government should allocate more resources in terms of capitation grants and fee bursaries to improve children's access to early childhood education in ASAL areas.

The Ministry of Education (MOE, 2006) in a letter for the launch of Nomadic Education Policy Framework expressed its role as those of enabling nomadic communities to realize the universal access to basic education and training. The ministry also emphasized collaboration of different stakeholders and the use of multifaceted approach in the delivery of education to the nomadic pastoralists. The results of this study also concurred with the views expressed above namely that of the collaboration of the different stakeholders and as well as the use of different approaches in increasing children's access to early childhood education.

CONCLUSION

The following conclusions were drawn from the study findings. First, several safety and security issues were preventing parents from enrolling their children in early years school in difficulty circumstance. The issues include Long distance from home to school, harsh weather, rugged terrain, salty water and ethnic conflicts.

Many strategies that could be used to address the challenges include establishment of mobile schools near villages, flexible school routine, avoiding rugged terrain when selecting sites for establishment of villages, boiling and filtering water for drinking and organizing frequent peace and reconciliation meetings.

RECOMMENDATIONS

Based on the findings of this study, the following recommendations were drawn for different key stakeholders:

Parents

1. Parents should ensure that all their children are enrolled in school. The findings of the study had shown that nomadic pastoralists did not enroll all their children in school specifically first and last born who were required to look after animals and participate in cultural ceremonies.
2. Parents should attend mobilization and education forums which aimed at capacity building. The parents should also be helped to understand educational needs of children and prioritize these needs in their pattern of movement so as to enable their children to access early childhood education. This is because the study had found that the community prioritized more the welfare of livestock that did not have much concern for the education of their children.

School Board of Management

1. They should organize education awareness forums for parents and other community members to help them understand the importance

of early years education. This is because the majority of young children in the area were not in school.
2. The school management should also work with many stakeholders to provide the required resources. The study had revealed that some children drop out of school due to lack of fees and other basic needs.

Ministry of Education, Science and Technology

1. The ministry should provide grants to children in schools in the community and fee bursaries and capitation grants towards other school costs. This will go a long way in enhancing children's access to early years education in the region. The support from the ministry will also facilitate recruitment and proper remuneration of pre-school teachers to boost their morale to teach and to make mobile schools achieve the target of enhancing access to education for these nomadic pastoralists.
2. The National Commission for Nomadic Education Kenya through the Ministry of Education, Science and Technology and the County Government should put in place local committees to help in ensuring that children in the nomadic pastoralist households are enrolled, retained and complete school. The local committees will also ensure the operationalization of Nomadic Education Policy Framework on the enrolment of children in early years education.
3. The National Commission for Nomadic Education in conjunction with Ministry of Education Science and Technology and the county government should put in place locational committees to help in ensuring that children of the nomadic pastoralists were enrolled, retained and complete school. The local committees' roles should be that of operationalizing nomadic education policy framework on the enrollment of children in early childhood education.

County Government

1. The county government should construct more schools. This is because long distances between the nomadic settlements and schools hindered children's access to early years education.
2. The county government should also spearhead mobilization and awareness for this community in order to empower them and enlighten them on the importance of early years education. The study had found that there was lack of concern on the part of parents on children's enrolment in school at the grassroots level.
3. The county government should support the establishment of more mobile schools to open access to education for children of nomadic pastoralists. The study had found that the existing mobile schools were not enough and some of them had closed down due to lack of teachers.

Faith Based and Non-Governmental Organizations

1. These organizations should facilitate community mobilization and education forums, monitoring of mobile schools, provision of resources and documentation of indigenous knowledge of these communities. The efforts by these stakeholders will contribute enormously towards children's access to early childhood education.
2. The Faith Based and Non-Governmental Organizations should also employ teachers. The study findings revealed had shown that teachers are poorly remunerated and shortages of teachers in early childhood centres. The study had also revealed that shortage of teachers can lead to closure of schools so there is need for adequate teachers in school.

REFERENCES

Abdi, A. I., (2010). *Education for all (EFA): Reaching nomadic communities* in Wajir, Kenya-challenges and opportunities (Doctoral dissertation, University of Birmingham United Kingdom).

Andrew, S. L., & Orodho, J. A., (2014). Socio-economic factors influencing pupils' access to education in informal settlements: A case of Kibera, Nairobi County, Kenya. *International Journal of Education and Research Volume 2*, Number 3.

Australian Early Development Census (2016). *Children developmentally vulnerable or at-risk on the Australian Early Development Census.* https://childatlas.telethonkids.org.au/cda-indicators/vulnerability-aedc-domains/.

Beatley, T., (2014). Urban design for a blue planet. In *blue urbanism* (pp. 61-83). Island Press/Center for Resource Economics.

Begi, N. (2015). Use of mother tongue as a language of instruction in early years of school to preserve the Kenyan. *Journal of Education and Practice, Vol. 5,* No. 3 2014.

Bongo, P. P. & Mudavanhu, C. (2015). *Children's Coping with Natural Disasters: Lessons from Floods and Droughts* in Muzarabani District, Zimbabwe. https://www.jstor.org/stable/pdf/10.7721/chilyoutenvi.25.3.0196.pdf?.

Bonds, S., (2012). *Food for Thoughts: Evaluating the impact of India's mid-day meal programs on educational attainment.* University of California, Berkeley.

Burde, D. & Linden, L. L. (2012). *The effect of village-based schools: Evidence from randomized controlled trial in Afghanistan.* https://www.nber.org/papers/w18039.pdf.

Convention on the Rights of the Child, G. A. Res. 44/25, U. N. GAOR, 44th Sess., U. N. Doc. A/Res/44/25 (1989). *Hague Convention on the Civil Aspects of International Child Abduction*, Oct. 25, 1980.

Duncan, G. J., Dowsett, C. J., Claessens, A., Magnuson, K., Huston, A. C., Klebanov, P., Pagani, L. S., Feinstein, L., Engel, M., Brooks-Gunn, J.,

& Sexton, H. (2007). 'School readiness and later achievement,' *Developmental Psychology*, vol. 43, no. 6, pp. 1428-1446.

Duze, C. O., (2010). Average distance travelled to school by primary and secondary school students in Nigeria and its effect on attendance. *An Interventional Multi-Disciplinary Journal, Ethiopia Vol. 4*. (4) Serial No. 17 pp. 378-388, October 2010.

Fernandez-Gimenez, M. E., Batkhishig, B., & Batbuyan, B., (2012). Cross-boundary and cross-level dynamics increase vulnerability to severe winter disasters (dzud) in Mongolia. *Global Environmental Change*, 22(4), 836-851.

Gakuria, A. R. (2013). *Natural resources based conflicts among pastoralists' communities in Kenya* (A project submitted to the University of Nairobi).

Gitau, D. N. (2013). *Impact of drought on primary schools learning in Laikipia West County*, Kenya, Kenyatta University, Unpublished Thesis.

Heckman, J. & Masterov, D. (2004). *The productivity argument for investing in young children.* Retrieved November 30, 2008 from http://jenni.uchicago.edu/Invest/.

Korkeala, O. K., (2011). *Climate and land in turmoil: Welfare impacts of extreme weather events and palm oil production expansion in Indonesia* (Doctoral dissertation, University of Sussex).

McDonald, N. C. (2005). *Children's travel: Pattern and influences.* VC Berkeley Earlier Faculty Research. https://escholarship.org/uc/item/51c9m01c.

McMillan, T. E. (2007). The relative influence of urban form on a child's travel mode to school. *Transportation Research Part A: Policy and Practice*, 41(1), 69-79.

Miako, J. K., (2012). *School levies and their effects on access and retention since the introduction of the subsidized secondary education* in Nyandarua North District, Kenya. Unpublished M. ED Thesis, Kenyatta University.

Migwi, S. M., (2010). *The impact of community support grant on access to early childhood education.* M. ED Thesis, University of Nairobi.

Ministry of education, (2006). *Nomadic Education Framework*, Nairobi Kenya.

Mulinge, C. N., (2012). *Factors affecting urban refugees in accessing education in public primary schools* in Kamukunji District Nairobi County Kenya (Doctoral dissertation, University of Nairobi, Kenya).

Panter, J., Jones, A., Sluijs E. V. &Griffin, S., (2011). The influence of distance to school on the associations between active commuting and physical activity. *Pediatric Exercise Sciences*, 2011: 23(1): 72-86.

Pence, A., & Shafer, J., (2006). Indigenous knowledge and early childhood development in Africa: The early childhood development virtual university. *Journal for education in international development*, 2(3).

Ramey C. T., Campbell, F. A., Burchinal, M., Skinner, M. L., Gardner, D. M. & Ramey, S. L. (2000). Persistent effects of early intervention on high-risk children and their mothers, *Applied Developmental Science*, 4, 2-14.

Ranabhat, M., (2014). *Determinants of access, participation and learning outcomes at primary level in Nepal* (Doctoral dissertation).

Republic of Kenya (2012). Children's Act. http://www.kenyalaw.org/kl/fileadmin/pdfdownloads/Acts/ChildrenAct_No8of2001.pdf.

Republic of Kenya (2013). Basic Education Act. http://ilo.org/dyn/natlex/docs/electronic/94495/117651/F-1505056566/KEN94495.pdf.

Republic of Kenya (2010). *The Constitution of Kenya*. http://extwprlegs1.fao.org/docs/pdf/ken127322.pdf.

Resa, R., (2007). *Factors affecting the enrolment and retention of students at a primary school in Anthra Pradesh-A village level study.* India.

Robson, C., (2004). *Real world research: A resource for social scientists and practiontioner- researchers* John Wilsey & Sons.

Ruto, S. J., Ongwenyi, Z. N., & Mugo, J. K. (2010). Educational marginalization in Northern Kenya. *Paper commissioned for the EFA Global Monitoring Report.*

Sanou, S., & Alkman S., (2005). Beyond access: Transforming policy and practice for gender equality in education. Pastoralists' schools in Mali. *Gendered roles and curriculum realities.*

Sengupta, S., & Ghosh, S., (2013). Poverty, child labour and access of schooling in India: Finding the gaps. *Asian Journal of Research in Social Sciences and Humanities*, 3(1), 203.

State of Early Learning in Australia (2017). *Early learning: Everyone Benefits.* Canberra, ACT: Early Childhood Australia. https://d3n8a8pro7vhmx.cloudfront.net/everyonebenefits/pages/73/attachments/original/1504689599/ELEB-Report-web.pdf?1504689599.

Sutter W. N., (2011). *Introduction to educational research.* Critical thinking approach. London: Sage Publisher.

UNESCO (2007). *EFA Global Monitoring report 2007: Strong foundations, early childhood care and education.* Paris: UNESCO Publishing.

Uwezo, (2011) *Are our children learning? Annual learning assessment report.* Nairobi. Georgebensons Media.

Vuri, D., (2008). The effect of availability and distance to school on children's time allocation in Ghana and Guatemala, understanding children's work project. *Working Paper Series.*

In: A Closer Look at Safety and Security
Editor: Jeff M. Holder

ISBN: 978-1-53618-176-0
© 2020 Nova Science Publishers, Inc.

Chapter 3

SAFETY, SECURITY, AND SOCIAL ENGINEERING — THOUGHTS, CHALLENGES AND A CONCEPT OF QUANTITATIVE RISK ASSESSMENT

Joachim Draeger[*]
IABG mbH, Ottobrunn, Germany

Abstract

Social engineering attacks on human operators supervising technical systems can pose a significant risk. Its quantitative assessment faces several challenges, however, which make the application of common assessment methods inappropriate. First, aspects of safety, security, and cognitive psychology have to be included concurrently. Second, the supervised technical system may possess a distinct inherent dynamics. Third, the available knowledge about the cognitive characteristics of the human operator will be limited. The complexity of the situation illustrated in the first point favors model-based considerations, whereby the limited knowledge indicated in the third point suggests the usage of comparatively simple models. As it turns out, the system dynamics paradigm seems to be a suitable modeling method to describe social engineering situations. This also takes into account the second point of the list, which leads to a preference for simulations to determine the outcome of specific scenarios.

---

[*] Corresponding Author's Email: draeger@iabg.de.

The representation of possible failures in the model together with their probabilities and criticality values is a prerequisite for a simulation-based calculation of the risk. Interpreting risk as mean expected criticality or, more generally, as probability distribution of outcome criticality leads immediately to a random sampling approach for covering the space of scenarios. The sampling is realized via Monte Carlo simulation runs. Missing knowledge can be integrated by variations of uncertain parameters according to a given probability distribution. Using the central limit theorem, the sampling error can be estimated and confidence intervals given.

For assuring the quality of the risk assessment, validation methods for the underlying model are discussed. Using the human driver of a semi-autonomous car as an exemplary application demonstrates the practical relevance of the considerations. The importance of *understanding* the risk assessment is substantiated as prerequisite for its credibility. The chapter closes with a short outlook, in which the proposed concept and opportunities for future research are discussed.

PACS: 02.50.Ng, 02.70.Uu

Keywords: cognition, attention, social engineering, risk assessment, Monte Carlo simulation, systems dynamics, semi-autonomous vehicles, malware

AMS Subject Classification: 62P15, 68U20, 92-08, 91E10

1. INTRODUCTION

The supervision of safety-critical processes — e.g., in the chemical industry, of distribution networks like power grids, or as a driver of a car — is still a human domain, at least as final authority. If a networked computer is involved e.g., as man-machine interface, a situation results subject to both safety problems and security threats. Social engineering attacks [1] are one possible threat. Greitzer et al. [2] give the following definition of social engineering:

> Social engineering, in the context of information security, is manipulation of people to get them to unwittingly perform actions that cause harm (or increase the probability of causing future harm) to the confidentiality, integrity, or availability of the organization's resources or assets, including information, information systems, or financial systems.

Accordingly, social engineering aims at a psychological manipulation of people into performing actions to the advantage of the attacker. As an example, an attacker may place a manipulated USB-stick [3,4] somewhere in the area of a company. An employee of the company passing by may pick it up and use it just for curiosity, infecting his computer with malware as a consequence. A large variety of other methods exists [5,6].

Social engineering is the most frequent attack method in IT security [2]. Several incidents have become well-known. Using a spear phishing email [7] as social engineering method, hackers attacked a German steel mill in 2014. They got access to the control systems of the production plant and produced significant damage [8]. As a second example, the Stuxnet worm [9] has compromised specific industrial control systems. Firewalls and air gaps were breached using infected USB-sticks as portrayed above. The malware switched off safety constraints. As a result, uranium enrichment facilities in Iran seemed to have suffered substantial damage.

Such incidents support the viewpoint that it is advisable to include social engineering and thus the human part [10–12] in risk assessments. Then, besides safety and security aspects, human factors and the domain of cognitive psychology have to be dealt with. Due to the interactions and trade-offs between all these domains, the risk assessment has to be executed in an integrated fashion. Isolated considerations for the various perspectives on the system are not adequate. Concerning safety and security, such a unified treatment is recommended in e.g., [13]. Examples of trade-offs between these two domains are given in [14,15]. Especially, industrial control systems like SCADA [16] have to coordinate security measures with real-time requirements of safety.

2. AIM

The complexity of the required considerations might be one reason, why the security-related study of human behavior and the ways of exploiting it is still insufficient [17]. As safety and security in earlier times [18, 19], psychological aspects and cyber security are typically discussed isolated from each other. The development of a psychologically well-founded risk assessment concept in social engineering situations would be a significant improvement, whereby the so-called risk measures the expected disadvantages associated with the usage of the system. Relying on *quantitative* risk assessments, as utilized for example in [20], additionally facilitates comparisons of system designs. The general perspectives of quantitative notions of risk are discussed in [21]. When executing a quantitative risk assessment of a social engineering situation, one has to pay special attention to two conflicting requirements. On the one hand, the representation of psychological aspects should be detailed enough for reproducing essential cognitive mechanisms, on the other hand it should be simple enough for assuring tractability of the intended risk assessment. In the following, the considerations are targeting at finding a reasonable compromise fulfilling both requirements.

For reasons of simplicity, we discuss social engineering only for interpersonal interactions with electronic means according to the taxonomy given in [2]. More precisely, we consider specifically a malevolent message aiming at an infection of the control software with malware causing harm to the technical system. Furthermore, a non-iterative (i.e. single-stage) social engineering attempt is assumed. We do not include the complete sequence of actions involved in the overall attack and its defence. Preparing and supporting actions like open source intelligence about the target [2] are omitted. Furthermore, we examine a single social engineering event only. Learning and training processes of the human operator are not taken into account. Such aspects — though viewed as important — are beyond the scope of this work. Dealing with the fundamental issues will already pose substantial challenges at the current state of knowledge.

The restriction of the modeling task to the concept level helps to focus on basic principles. This takes into account that in the domain of cognitive psychology details are still unclear and concurrent theories of explanation exist [22]. Inconsistencies in various studies are indicated e.g., in [2]. This is an obstacle, because a suitable representation of cognitive psychology is essential for reliably predicting the success rate of social engineering attacks.

One may be tempted to ask, why safety aspects are included in the following investigation of a social engineering situation. Two reasons are responsible for this decision. First, the safety part of the overall system defines the criticality of the consequences of security violations. In the given situation, a security violation means an infection of the supervised technical system with malware. The resulting damage is a measure of criticality. The domains of cyber security and psychology alone do not allow an evaluation of the consequences in a malware infection scenario. Second, a model describing only the human interaction with a social engineering attempt may give an inadequate picture. Many of such attacks are successful despite of the threat known to the victim, because his mind is insufficiently focused on the attempt leading to a poor situation awareness, a wrong perception of the situation, or the realization of other cognitive weaknesses [23, 24]. A quantification of focus and attention seems to be only possible, however, if some aspects of the cognitive tasks concurrently active at the moment of the social engineering attempt are known.

The chapter is organized as follows. Section 3 presents related work. The basic requirements of a model usable for a simulation-based risk assessment of social engineering situations are discussed in section 4. As it turns out, a great challenge is the discrepancy between the complexity of the cognitive processes of a real human being and their necessarily simplified representation in the model used for the risk assessment task. Accordingly, the cognitive aspects are elaborated in more detail in section 5. The risk assessment by Monte Carlo simulation runs, which is based on the model developed in the previous two sections, is presented in section 6. Section 7 gives some remarks about validation aspects. Especially, options for building confidence in the model and in the risk determination process are discussed. Semi-autonomous cars are discussed as an application in section 8. The chapter closes with an outlook in section 8, which presents some opportunities for future research.

3. RELATED WORK

The examination of methods for a simulation-based risk assessment of social engineering situations is an interdisciplinary project. Many different science domains and subject areas are involved. For clarity, related work is thus referenced in situ, when the corresponding topic emerges. At this point, only some general surveys are indicated for readers interested in comprehensive background information.

The large number of social engineering attacks has led to many general overviews, e.g., [2, 6, 24, 25]. The psychological perspective is elaborated in [26]. Several models of a social engineering situation have been created, see for example [5, 27, 28]; these models are not aiming at enabling quantitative risk assessments, though, and are not very detailed from the psychological perspective. Overviews about existing approaches for modeling cognitive aspects of human behavior are given in e.g., [22, 29]. Several papers are dedicated specifically to the representation of the cognitive processes involved in the control of dynamic systems [30, 31].

Integrated risk assessments for both safety and security are still not very common, though their popularity is increasing rapidly. We thus provide some general references about this topic. Several authors (e.g., [18,19]) stress the possibility of a unified risk assessment for both safety and security despite of significant differences between these two domains [32–35]. An overview of methods, which may be usable as a common framework, is given in [36]. Notions of risk, which are usable for both safety and security, are considered in [37, 38].

4. SIMULATION-BASED RISK ASSESSMENTS

4.1. Basic Modeling Aspects

It is doubtful, whether discussions at a purely informal level will suffice for executing adequate risk assessments in the case of complex systems [39]. Specifically in the context of socio-technical systems [40,41], deficits regarding their quality were found to be quite frequent. The observed bias seems to be clear. As soon as humans with their overwhelming behavioral complexity are included, profound insight into the system and its internal relationships is usually missing.

The problems of informal approaches are arguments for the use of formally oriented methods [42]. As a consequence we will take a model-based viewpoint. The usefulness of such an approach for risk-related considerations is demonstrated in [43, 44]. A survey of various model-based risk assessment methods for the specific case of systems control with inclusion of both safety and security can be found in [45].

The dynamics of the technical system supervised by the human operator may be inherently non-trivial. Accordingly, neglecting system dynamics would be an oversimplification [46]. The dynamically changing state of the supervised technical system influences, at which times the human operator will deal with a social engineering attempt and how successful this attempt will be. If the attempt is not accepted or declined immediately, the human operator will start to explore the information associated with the attempt. This is a dynamical process as well as the ongoing competition for attention with other cognitive tasks. Finally, in the case of a malware infection, faults and failures may dynamically propagate across the technical system. New faults may arise and interfering with problem mitigation tasks dealing with other faults. All these effects make it adequate to include the overall dynamics in the assessment of the risks [47]. A simulation-based risk assessment is thus considered as necessary.

4.2. Modeling Paradigm

Socio-technical systems [48,49] are necessarily complex systems [50], since the complexity of human behavior dominates the complexity of the overall system. Notwithstanding, the system model has to be sufficiently simple for assuring fast simulations of the system model. A short run-time is necessary for processing the large number of simulation runs, which are required for covering the variations of e.g., human personality, experience, and form of the day as we will see later in section 6.2. As a consequence, significant abstractions will be involved when transiting from the real system to its representation in a model.

The inaccuracies resulting from these abstractions seem to be readily acceptable as the involvement of humans makes the risk assessment task ill-defined anyway. Precise predictions cannot be expected. Furthermore, more detailed models will usually also be accompanied with a higher number of model parameters. The additional knowledge about suitable parameter values may hardly or not be accessible at all, however. It is not plausible to assume comprehensive knowledge about e.g., the cognitive characteristics of an individual human operator. We have to take the trade-off between model accuracy and minimization of uncertainties into account. This means, that choosing a comparatively simple model is not necessarily a bad choice.

For dealing with complex systems at a high abstraction level [51], the system dynamics paradigm developed by J. Forrester is a natural option [52]. We are following [53] for a short introduction. Basically, a system dynamics model consists of stocks and flows. The stocks represent the (continuous) state of the system, which are modified by the flows between the stocks. A flow f from the stock s to the stock s' increases the value of s and correspondingly decreases that of s'. A system dynamics model consisting of stocks $s_1, \ldots, s_k$ is equivalent to an ordinary differential equation system over the variables $s_1, \ldots, s_k$. This gives the model a precise mathematical meaning [54]. In this context, the flow f means

$$\frac{ds}{dt} = \cdots - f \cdots$$
$$\frac{ds'}{dt} = \cdots + f \cdots \qquad (1)$$

Since differential equations are used throughout all science domains, system dynamics is an interdisciplinary approach not least demonstrated in the world model discussed by the Club of Rome [55]. This generality is crucial, because in our case the model has to represent aspects of safety, security, psychology, and the technical system being supervised simultaneously. Differential equations can be solved efficiently using standard methods of computational mathematics. A high simulation speed can thus be guaranteed. Sometimes, the system dynamics methodology is enriched by events and other features [56].

System dynamics models have already been successfully used for representing complex systems from the safety and security perspective at high abstraction level. As an example, one may look at [57]; this reference also contains considerations about the closeness of the model to reality in view of the strong abstraction. Model-based discussions of the insider threat of cyber security using the system dynamics paradigm can be found in e.g., [58–61]. Several papers [60, 62–64] are dedicated to qualitative evaluations of risk posed by insider threats. Greitzer et al. [2] have represented a social engineering situation by a system dynamics model. The model is not detailed enough for determining the criticality of specific situations, however. Cai and coworkers [65] provided a general cognitive behavior model based on the system dynamics modeling paradigm. Taking together these contributions, it seems feasible to use system dynamics for providing a psychologically well-founded model of a social engineering situation.

4.3. Model Structure

A model M of a social engineering situation as described in section 2 consists of three principal components (see figure 1). The first component M_T represents the safety-critical technical system supervised by the human operator. The second component M_S describes the social engineering attempt in form of e.g., a message. The third component is the human operator M_C himself controlling the two other components M_T and M_S. In this section, the internal function of M_T and M_S is discussed. Their internal structure is graphically depicted in the lower part of figure 2. Since these components are belonging to the technical resp. computer science domain, their models can be considered as quite conventional. Accordingly, we will focus on their interplay with M_C.

The technical system M_T is composed of a software subsystem susceptible to a malware infection and a physical subsystem determining the severity of various malfunctions and thus the level of their maybe disadvantageous consequences. An infection with malware by M_S will degrade the capability of the software subsystem to control the physical subsystem. This means, that a malware infection will increase the fraction of failures of the physical subsystem.

Safety, Security, and Social Engineering

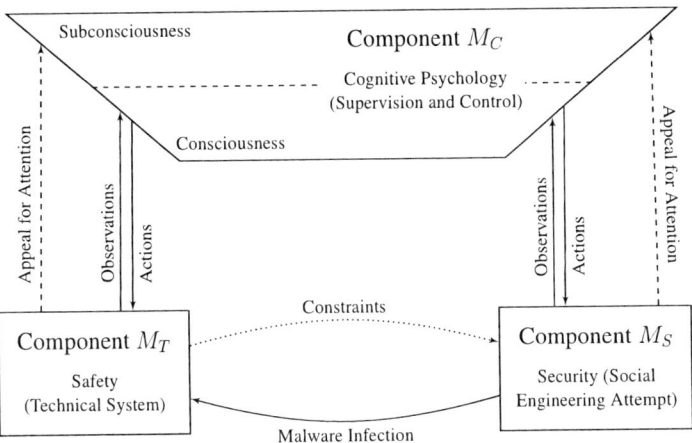

Figure 1. Basic structure of a simulation model usable for risk assessment purposes.

The coupling of the technical system M_T to the cognitive component M_C, which constitutes the human operator, is realized by a representation of the man-machine interface MMI_T, which is located between the software subsystem of M_T and the cognitive component M_C. It translates the state of the technical system to a state representation, which can be processed by the cognitive component M_C. Exemplary variables provided by MMI_T and serving as input for M_C may be the amount of accessible information about the technical system, the level of expertise required for dealing with the current situation, the complexity of pending decisions, the level of completeness of the available information and the fraction of faulty information. As a counterpart, the output of M_C to the technical system may include the number and the accuracy of actions required by the human operator.

The man-machine interface MMI_T interprets the state of the technical system from the perspective of the cognitive demands, which the human operator are facing in this situation. At the side of the cognitive component M_C, the description of the cognitive demands is matched with the available cognitive capabilities and resources. Depending on the outcome of the match, the current quality of the human control performance is predicted. This quality is propa-

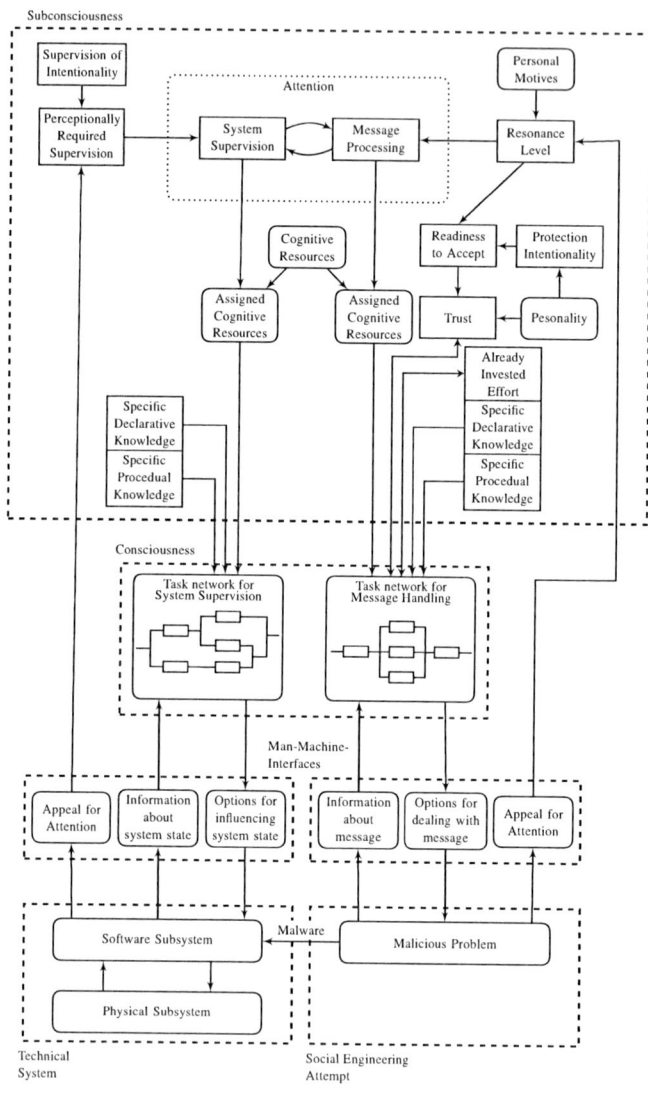

Figure 2. Details of the modeling concept presented in figure 1 showing the internal mechanisms and processes. Boxes with rounded corners indicate modeling objects, which are not elementary stocks but larger units composed of stocks and flows between them.

gated from the cognitive component M_C via MMI$_T$ to the technical system M_T. In M_T, the quality of the human supervision task influences the probability (i.e. frequency) of a disadvantageous system evolution. A malware infection of the software subsystem of M_T will usually also degrade the function of the man-machine interface MMI$_T$ e.g., concerning completeness and correctness of the mediated information.

With regard to the component M_S we assume, that an incoming message comprises the social engineering attempt. In the case of M_S, the control task executed by the human operator does not consist of a continuous supervision. Instead, the topic of the mail message determines whether he is principally interested or not. If he is interested and clearly trusts the message, he will immediately accept the offer provided by the message. Without interest or trust, he will immediately decline. As long as he is unsure about the trustworthiness of the message, he will further explore the information provided by the mail message as long as his interest keeps up with the already invested effort.

Analogously to M_T, we have to provide some kind of man-machine interface MMI$_S$ responsible for providing a representation of the message accessible to the cognitive component M_C. The input variables of M_C made available by MMI$_S$ may consist of e.g., the message topic, its degree of expressiveness, the amount of available information, and the degree of consistency of available information. Possible outputs are the actions of accepting the provided offer, declining it, and further exploration like gathering background information. Accepting the offer will in our case trigger the malware infection of the software subsystem of M_T.

5. COGNITIVE COMPONENT

5.1. Overall Architecture

The success of a social engineering attempt can have many different reasons ranging from perception errors over wrong decisions to inadequate actions. Accordingly, the cognitive component M_C, which models the behavior of the human operator, must have a broad coverage. This follows Newell's principle of unified cognition [66]. Since predictive purposes are intended, the model has to be based on cognitive principles instead of just giving a phenomenological description [67]. The papers [22, 68–72] provide surveys of methods representing the human behavior with emphasis on cognitive aspects. A thorough inspection of capabilities and characterizing properties led to the conclusion, that no existing model known to the author seems to be adequate for dealing with the intended application. This may be astonishing facing the large number of initiatives to provide full-scale cognitive models. In many respects, cognitive models are still at the level of fundamental research though. The specific requirements posed by an assessment of social engineering risks were not in the focus of their development. For some models, the coverage of the psychological domain seems to be insufficient — both rational behavior for the supervision task and emotional aspects for the susceptibility concerning the social engineering attempt have to be included. Other models are agent-oriented aiming primarily at representing interactions between humans. Rule-based cognitive architectures will face severe runtime problems due to a very detailed description of cognitive aspects.

As a consequence, the cognitive component M_C uses the system dynamics paradigm as modeling concept, which can cover all required issues. It allows a fast execution of many simulation runs enabling the intended risk assessment. The structure of M_C — shown in the upper part of figure 2 — is determined by the requirement to incorporate all aspects of cognition, which seem to be essential for predicting the behavior of a human operator confronted with a social engineering attempt. Due to [73], both non-rational and rational issues have to be included. Whereas rational issues like the available knowledge essentially determine the principal cognitive capabilities, the non-rational issues will influence up to which amount these capabilities are actually applied. As an example, the intentionality of the human operator to protect the technical system from harm is a major non-rational influence. Whereas the rational issues are mainly associated with the consciousness, the non-rational issues seem to be mainly localized in the subconsciousness.

The manifold factors modulating the cognitive performance are not covered here, though they may have a considerable influence on the cognitive performance. A person being in fear [74], for example, more readily violates rules. Different lists of modulating factors are given in e.g., [2, 75, 76]. Up to now, it seems to be impossible to identify the most important factors. The number of well-documented examples of social engineering is insufficient for drawing reliable conclusions in this respect [2]. The limited number of detailed case investigations impedes the calibratability of cognitive models. It restricts the precision, which is achievable by their simulation. This does not compromise the practical usability of our considerations in principle, however. One just cannot expect the derivability of globally valid, simple statements of mathematical exactness.

5.2. Cognitive Processes

Due to the limited expressiveness of system dynamics models, it is recommended to adapt the cognitive component M_C to an explicit description of the executed cognitive tasks instead of using a correspondingly configured generic model. This gives M_C a process-like structure. It has the advantage that phenomena like fault propagation can be naturally represented.

According to this modeling strategy, we have to represent the cognitive processes associated with the supervision of the technical system and with the handling of the social engineering attempt explicitly. This can be reached in a straightforward way by decomposing them in sub-tasks related to observations, decisions, and actions [31]. If necessary, one may adapt the decomposition to the peculiarities of the represented control task or choose a more detailed decomposition. This resembles the Queueing Network-Model Human Processor (QN-MHP), which is a cognitive architecture specifically designed for analyzing the multitasking performance in human-machine systems [77–80]. Here, the term 'multitasking' is used in an informal sense. The question, to which extent humans are capable of 'true' multitasking, is still subject to intense debate. In the case of social engineering, the two concurrent tasks are not executed simultaneously with approximately equal effort. Rather, the involved person focuses on one task actively, though a rudimentary monitoring may very well happen concerning the other. Based on this comprehension, the multitasking perspective can be modeled [72] by introducing the concept of attention. Depending on the focus of attention, the majority of cognitive resources (see section 5.3) is assigned to one control task or the other. The concept of attention is discussed in section 5.4.

A refinement of the task networks sketched above consists in breaking down the decision task according to the so-called dual process theory [81, 82]. The duality of human decision processes seems to be the result of an evolutionary optimization of human decision making with respect to effectiveness on the one hand and the cost of cognitive biases and of many different sources of potential errors on the other [83]. The two decision processes are a fast error-prone satisficing approach and a slow accurate rational process. The satisficing approach aims at an unconscious recognition of known patterns, whereas the latter generates consciously a variety of options and selects the best one.

Following [84], the fast satisficing approach — also called recognition-primed approach — is applied, if the following conditions hold: low perceived problem complexity, available procedural knowledge and adequate cognitive resources. Its usage [17,85], which regretfully accepts systematic errors, becomes plausible as soon as more rational solutions overstrain the available cognitive resources and/or violate physical constraints [86] like time limits. If these conditions are not fulfilled, the rational approach is applied. It is recommended to take the dual process theory into account, because the two types of decision processes differ considerably concerning frequency of errors and consumption of cognitive resources. In this way, significantly more accurate risk assessments can be realized. Furthermore, social engineering is an attack method essentially trying to exploit the biases and imperfections of the satisficing mode [87]. These points of attack do not exist in the rational mode. The less prominent the rational mode, the more promising is a social engineering attempt trying to exploit the indicated weaknesses [5].

5.3. Cognitive Resources

The cognitive resources, available to the human operator, are restricted [88–90]. For example, knowledge, reasoning capabilities, perceived (and real) time bounds for providing a solution, and the capability to deal with complexity significantly influence the quality of the cognitive processes and should thus be a part of the model component M_C. Their respective amount can be represented in M_C as stock levels. The levels of available cognitive resources can be matched with the corresponding levels of cognitive demands posed by the current state of the technical system or the social engineering attempt, which are provided by MMI_T and MMI_S. If for example the complexity of the current situation is considerably larger than the cognitive capability to deal with complexity, the probability of an erroneous decision will raise significantly. If the operator is not able to grasp the large amount of information in a message, cues indicating a social engineering attempt may be overlooked [2].

The various kinds of knowledge [91] are some of the most important cognitive resources. With respect to the dual-process theory it is recommendable to distinguish explicit declarative knowledge and implicit procedural knowledge [22,92]. Procedural knowledge may consist for example of preconceived opinions. Such knowledge may be applied successfully despite low attention, contrary to explicit knowledge. Thus, an experienced human operator may focus his attention on the social engineering attempt while still reaching an acceptable quality of supervision of the technical system. This will not be possible for an inexperienced operator. Additionally, we have to distinguish between knowledge related to system supervision and knowledge related to social engineering.

The boundedness of cognitive resources has far-reaching consequences. One implication is that every action is associated with cognitive costs. Even the acquisition of information consumes resources (time, energy, attention etc.). Accordingly, the human operator will only be able to make a limited number of observations in a given time interval. Thus, the limitation of cognitive resources may lead to uncertainties and thus necessarily to suboptimal decisions. Following Griffiths et al. (e.g., [93] and references therein to similar approaches), the boundedness of resources is responsible for the observed bounded rationality of humans in general. The higher the level of usage of the available cognitive resources, the higher the perceived cognitive workload [84], which will be noticed as some kind of stress. Stress is one of the factors modulating cognitive performance, but the relationship is rather non-trivial. At low levels, stress can be advantageous e.g., due to reduced response times. Only after reaching a certain threshold, stress becomes disorganising, whereby the threshold (the so-called stress-resistance) depends on personality, experience, and numerous other factors [84]. Concerning dual-process theory, high stress levels seem to trigger a preference of the satisficing decision approach. This would be a plausible mechanism of the brain for reducing the cognitive workload and thus the stress level. As a consequence, at high stress levels the human operator may not be willing to check the internal consistency of information. He will thus easier become a victim of a social engineering attempt.

5.4. Attention and Multitasking

The human operator is facing a multitasking problem. He has to 'control' the two components M_T and M_S concurrently. As a consequence, he has to focus his attention on one of the components depending on the current situation and the existing motives. In this context, a motive is a psychological force, which drives a human to an action [94]. Motives determine the readiness of an individual to focus the cognitive capabilities to a specific task, which in turn will get also the majority of the available cognitive resources [84]. Only a small part of the cognitive resources will remain for other tasks.

The focus of attention is controlled by the subconsciousness. Concerning the general interplay between conscious and sub-conscious processes, we are following the layered reference model of the brain [95–97] and the cognitive model used in PECS [94, 98]. Other approaches integrating these two components of the human mind in a unified model are presented for example in [65, 99–101].

One motive of the human operator is the intentionality to supervise the technical system. The intentionality of supervision together with the required level of supervision, which results from the current state of the technical system, controls the level of attention towards the technical system. The social engineering attempt is competing for attention by appealing other motives. Here, just some examples of such motives are given. The perception of the originator of the social engineering attempt as a likeable person may one of the reasons to become a victim of the attack [102]. As some kind of a counterpart, humans have a natural desire to be liked by others; accordingly, they are tempted to some extent to do other people a favor [17]. The motive of prosocial behavior [103] is strong enough to violate security policies as well. Other motives are curiosity, desire of excitement, greed, and so on.

The various motives of the human operator are determined by his personality [104]. In the case of the social engineering attempt, the topic of the message gives the motives being addressed by the message. The human operator will only be interested in the message (being in fact a social engineering attempt), if the message topic appeals to one of his personal motives. The stronger the resonance between message topic and personal motives, the stronger the readiness to accept the offer provided by the message. For switching his attention from the supervision task to the social engineering attempt, the resonance between motives and message topic has to be strong enough for surpassing the motive of intentionality to supervise the technical system.

A strong resonance with a motive will usually not trigger an immediate acceptance of the offer provided in the message. It exists control mechanisms. One of these mechanisms is trust [105]. If the human operator does not trust the incoming message, the social engineering attempt will usually not be successful [102, 106]. Trust results from the feeling that the assertions posed in the message are true, and this in turn may be the consequence of many consistent data. The usual way of thinking is as follows: The higher the number of consistency tests turning out to be true, the higher the probability that the data themselves are true. Accordingly, a more detailed story seems to be more of consistency tests turning out to be true, the higher the probability that the data themselves are true. Accordingly, a more detailed story seems to be more credible. The higher the effort invested in elaborating the internal and external consistency of a social engineering message, the higher the generated trust. If the trust is only small (i.e. if the message looks suspicious to the human operator), transiting to the rational mode is preferred under the condition of sufficient cognitive resources. Since the rational mode requires more cognitive resources than the satisficing mode, humans tend to prefer to trust another person per default instead of doing checks and re-checks. This principle of economy affects the control of motives also in another way. If the human operator has already invested a significant amount of cognitive resources in the exploration and analysis of the social engineering attempt without reaching a decision, resonance with motives will usually decline. This reflects some kind of cost-benefit tradeoff. Additionally, a risk-averse personality is counteracting the impulse to fulfill the motives triggered by the social engineering attempt.

5.5. Resulting Big Picture

The preceding sections provide details, which have to be included in the cognitive component M_C for realistically reproducing the behavior of humans facing a social engineering attempt. Putting these details together, we get the following big picture of the internal function of the cognitive component M_C (see figure 2).

The component M_C consists of two subcomponents, the consciousness and the subconsciousness. Task networks modeling high-level cognitive processes comprise the consciousness subcomponent. The operation of these task networks essentially depends on the assigned cognitive resources, which are in turn determined by the focus of attention. Subject to the motivational response triggered by the technical system M_T and by the social engineering attempt M_S, the attention of the human operator is directed either towards M_T or M_S, closing the loop in this way. Both cognitive resources and the focus of attention are controlled by the subconsciousness, which thus has a significant influence on the operation of the high-level cognitive processes in the consciousness.

6. Concept of Risk Assessment

6.1. Errors and Risk

Usually, system models will represent only the nominal behavior of a system. A model for risk assessment purposes has to include off-nominal modes of the system as well, however. Accordingly, fault and failure mechanisms have to be represented in the model M.

The state of a system dynamics model consisting of stocks $s_1, \ldots, s_N$ is given by the tupel $(s_1, \ldots, s_N)$. The flows between the stocks are changing the state. If we additionally assume the conservation law $\sum_i s_i = 1$, a reinterpretation of the flows as state transitions becomes possible. Now, the level of stock s_i may represent the frequency (i.e. probability) of the occurrence of the state associated with s_i. Staying in this picture, an extension by error states is easily possible. One just has to supplement the stocks corr_i representing nominal system states by stocks error_j representing off-nominal states. Choosing suitable initial conditions and (re)scaling the flows appropriately will lead from $\sum_i \text{corr}_i = 1$ to $\sum_i \text{corr}_i + \sum_j \text{error}_j = 1$. Modulating factors like stress or additional knowledge will modify the probability of the occurrence of error states by changing the flows — i.e. the frequency of the state transitions — between the stocks.

For the state transitions in a component or for the state propagation from one component to another, different transition types can be distinguished. In figure 3, the state propagation process from an elementary cognitive task T to an elementary task T' is shown. Via such propagation mechanisms, errors occurring in a specific step of a cognitive or technical process may propagate throughout the whole process. The frequency of error states may sometimes increase from one process step to the next due to the occurrence of new errors and sometimes decrease due to fault recovery mechanisms. Cognitive errors of the human operator may propagate to the technical system component M_T in course of the human control tasks. The errors occurring over time will finally determine the criticality value assigned to the outcome of the system evolution.

Typically, the component M_S will not contain error states in the usual sense. This is partially due to the fact that the basic purpose of the social engineering attempt is to provoke just such errors in the component M_T, provided that the human falls victim to the attempt. Instead, the states of the component M_S indicate inconsistencies in the information provided by the social engineering

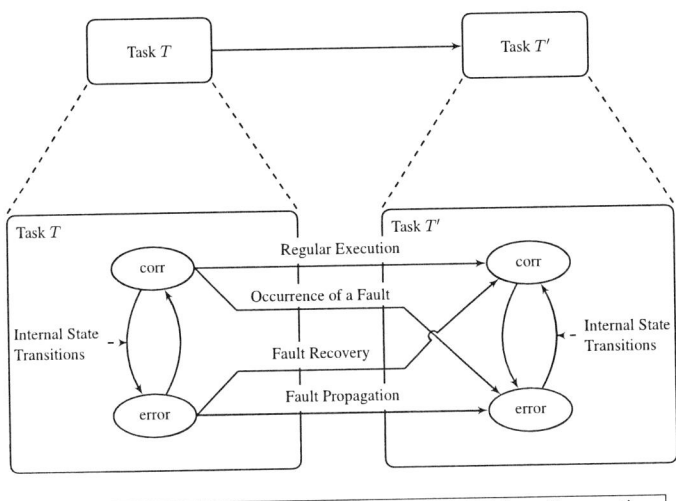

Figure 3. State propagation from a task T to a task T' in a cognitive task network, whereby T, T' have two internal states corr, error each. Increasing the number of internal states will give of course more options of state propagation.

message, which can be uncovered by the actions of the human operator exploring the details of the message and of associated background information. Any such inconsistency will decrease the trust in the offer provided by the message, eventually leading to its denial or further actions of exploration. This does not mean, however, that the exploration process itself will be free of errors. Though the component M_S will usually not contain error states itself, the cognitive task network belonging to the consciousness part of M_C and dealing with M_S may be subject to processing errors very well. Information may be overlooked, aspects of the message may be perceived in the wrong way, an erroneous click on a link contained in the message may be made, and so on. Quality and quantity of errors in the cognitive task networks are significantly influenced by the subconsciousness via the assigned cognitive resources. Analogously to M_S, the

Table 1. A detailed error classification scheme

Error Type	Representation in Model
Excessive usage of cognitive resources	Increased workload
Absent activities, missing reaction	Correspondingly modified frequency distribution of disadvantageous outcomes
Unnecessary activities, superfluous reaction	Correspondingly modified frequency distribution of disadvantageous outcomes
Incorrect activities	Correspondingly modified frequency distribution of disadvantageous outcomes
Wrong timing of activities	Time pressure or control failure
Incorrectly orchestrated activities (e.g., wrong order of execution)	Rate of errors increases with number of activities
Correct activities	Optimal control performance

subconsciousness part of the cognitive component M_C does not contain error states either — it contributes to the human weaknesses by bounding human rationality.

The higher the number of different error states, the finer the possible resolution of error effects. The error classification scheme presented in table 1 is based on the error categorization given by Daniel Scherer et al. [107] from the perspective of electrical systems operations. Here, it is considered as an adequate starting point for general systems as well. Examples of possible extensions are shown in figure 4. It results a modeling concept, which can reproduce a variety of error types covering the technical system, cognitive errors, and multi-tasking related errors and biases resulting from limited cognitive resources.

Adopting the state-transition picture of system dynamics as described above, a simulation of the system model M gives the probability — i.e., frequency — $p(q)$ of possible outcome states $q \in Q$ for the chosen model parameters. Additionally, the part of the outcome state belonging to the physical

Safety, Security, and Social Engineering

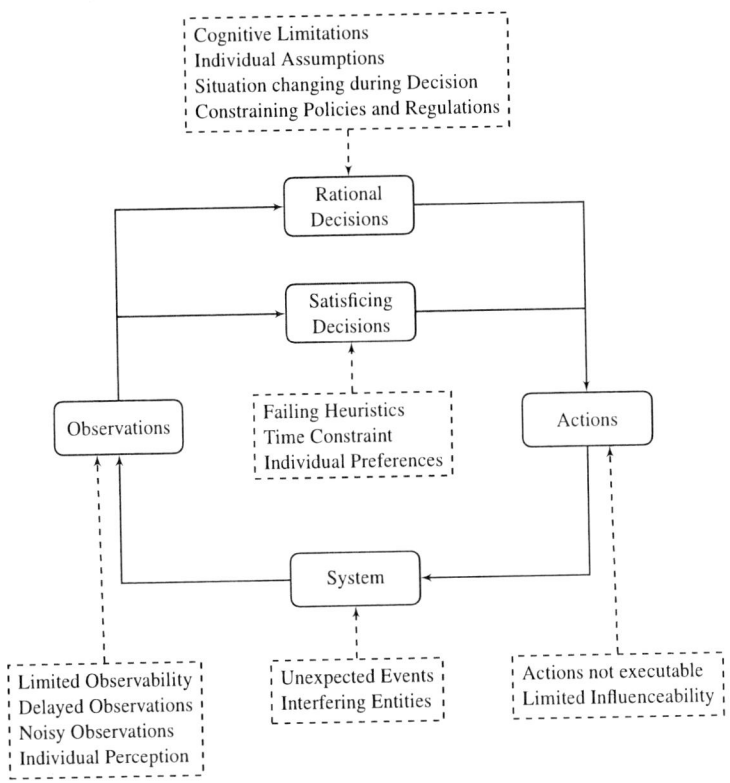

Figure 4. Some advanced types of cognitive errors, which may influence the control tasks of the human operator.

subsystem provides the criticality values $c(q)$ of the outcomes q, which measure the amount of disadvantages associated with their realization. Taken together, the probability $p(q)$ of occurrence of an outcome q and its assigned criticality $c(q)$ determine the risk contribution $R(q) = p(q) \cdot c(q)$ of the outcome q. An aggregation of the risk contributions $R(q)$ over all possible outcomes $q \in Q$ gives the desired overall risk $R = \sum_{q \in Q} p(q) c(q)$. This procedure reflects the usual approach [108] of determining the risk R. Details of such a simulation-based notion of risk can be found in [109, 110].

6.2. Inclusion of Uncertainties

For a social engineering situation, usually a variety of uncertainties exists. The personality of the human operator and his cognitive capabilities resp. resources belong to the most important sources of uncertainties. They can have a significant influence on the outcome of a social engineering attempt [2, 31]. Typically, a high level of awareness decreases susceptibility against social engineering attempts [111]. High levels of openness [112], obedience (the readiness to follow instructions) [17], normative commitment, or trust [113] will usually increase the susceptibility. Other important influences are the attitude towards work [73] and the perception of risk (underestimation of the quality of attacks and/or overestimation of the quality of defense options).

Similar to the human operator, the characteristics of the attacker and thus also of a future social engineering attempt may be subject to large uncertainties. It will usually be launched with skills, resources and motivations unknown to the human operator and the analyst tasked with the risk assessment [35]. Other sources of uncertainty are the cognitive state of the human operator in general — the performance of humans may vary due to his state of exhaustion, eventual medical problems etc. — and the state of the technical system at the time of the social engineering attempt.

For dealing with these uncertainties, we replace a single scenario uniquely characterizing a given situation by a whole set of scenarios corresponding to the variations of the uncertain parameters. As common for risk assessment purposes, we quantify uncertainties by probability distributions [114, 115]. The description of parameter variations by probability theory has the additional advantage that it can be perfectly integrated with the probability-based notion of risk. The inclusion of uncertainties in the safety and security context is discussed in detail e.g., in [116–118].

6.3. Sampling-Based Calculation

The state-transition picture of the system dynamics paradigm developed in section 6.1 allows an efficient simulation-based calculation of risk contributions. Due to the many uncertain model parameters as indicated in the preceding section, notwithstanding an enormous number of simulation runs may have to be executed for determining the resulting risk within a given error limit. This emphasizes the importance of providing a simple cognitive model, which can be simulated at high speed.

The uncertainties typically admit an infinite number of different parameter values. Thus, it is not possible to include the complete set of admissible scenarios in the risk assessment. An approximative approach has to be used instead, which includes only a finite number of scenarios. This finite subset of possible scenarios may be selected randomly based on a so-called Monte Carlo approach. In the following, its basic aspects are repeated. The Monte Carlo method has been successfully applied to cyber security-related risks in e.g., [119–121] and to safety in e.g., [122, 123]. An application to risks associated with a malware epidemics in a networked embedded system can be found in [57].

Let us assume that the criticality value $x = c(q)$ in the risk measure $R = \sum_q p(q)c(q)$ is distributed with a frequency (i.e., probability) distribution $f_c(x)$. Accordingly, it holds

$$R = \sum_q p(q)c(q) = \sum_x x f_c(x) = \mathsf{E}(c) \tag{2}$$

whereby the operator E designates the expected value. Based on this interpretation, we can formulate an estimate of the risk R based on a finite subset $\bar{Q} \subset Q$ of the set Q of all outcomes. The finite subset $\bar{Q}$ can be provided as random sample $X_1, \ldots, X_N$ of outcomes distributed as c, i.e., following the probability distribution $f_c(x)$. If c is integrable, i.e, if $\mathsf{E}(c)$ indeed exists for the probability distribution f_c, then the sample average

$$Z_N = \frac{1}{N} \sum_{i=1}^{N} X_i \tag{3}$$

is an unbiased empirical estimator of the expected value $\mathsf{E}(c) = \mathsf{E}[X_i]$ [124]. This gives $Z_N \approx \mathsf{E}[X_i] = \mathsf{E}(c)$ because of the linearity of the expected value operator. More precisely, it holds

$$Z_N \xrightarrow{\text{almost sure}} \mathsf{E}[X_i] = \mathsf{E}(c) \tag{4}$$

due to the strong law of large numbers [124].

Such a random sampling approach for an approximative risk assessment will only work, if the set $\bar{Q}$ of randomly selected outcomes is indeed representative for the set Q of all outcomes. Otherwise, the calculated risk value may be no good approximation of the exact value. For granting the required representativeness, it may suffice e.g., to exclude the existence of rare events with high criticality. Such so-called Low-Probability High-Consequence (LPHC) events have an exceptionally large influence on the overall risk R according to their definition [125], but their small probability will lead to their inclusion in $\bar{Q}$ only occasionally.

A quantitative discussion of the approximation error of a random sampling approach is possible using the central limit theorem. If both the expected value $\mu = \mathsf{E}(X_i)$ and the standard deviation $\sigma = \sqrt{\mathsf{Var}(X_i)}$ exist, the central limit theorem is applicable and gives

$$Z_N \xrightarrow[N \to \infty]{\text{distribution}} \mathcal{N}\left(\mu, \frac{\sigma^2}{N}\right) \qquad (5)$$

whereby the notion $\mathcal{N}$ designates the standard distribution.

The central limit theorem leads to an important statement concerning the convergence rate of the Monte Carlo approach. Designating the sampling error as $\varepsilon_N := Z_N - \mu$, we can state that $\sqrt{N}\varepsilon_N/\sigma$ converges in distribution to a Gaussian random variable with zero mean and variance equal to one. Increasing the number of Monte Carlo runs will thus decrease the approximation error, whereby the convergence rate is rather slow ($1/\sqrt{N}$). As an immediate consequence, it additionally holds

$$\lim_{N \to \infty} P\left(|\varepsilon_N| \le \alpha \frac{\sigma}{\sqrt{N}}\right) = \frac{1}{\sqrt{2\pi}} \int_{-\alpha}^{\alpha} \exp\left(\frac{-x^2}{2}\right) \mathrm{d}x =: \beta \qquad (6)$$

for all $\alpha > 0$. For large N, the expected value $\mathsf{E}(c)$ will belong to the confidence interval

$$\left[Z_N - \alpha \frac{\sigma}{\sqrt{N}}, Z_N + \alpha \frac{\sigma}{\sqrt{N}}\right] \qquad (7)$$

with probability β [126]. The confidence interval given in (7) quantifies the approximation error in a Monte Carlo approach.

7. VALIDITY ASPECTS

A validation [127] of the simulation model is mandatory for assuring the quality of the risk assessment. The model is provided by human intervention, and it may not be necessarily adequate due to the inability of humans to grasp the situation in its entire complexity. An inadequate model may lead to risk values, which will deviate from reality. The model validation has to be supplemented by showing the correctness of the calculation of risk based on the given model. A general overview about the validation of cognitive aspects is given in [41]. Johannsen et al. [128] discuss the validation of socio-technical systems.

Some peculiarities of social engineering are serious obstacles for a validation. First of all, an ill-defined situation is considered, last but not least due to manifold psychological aspects. Furthermore, the extent of abstractions contained in the model limits the similarities between the computational model and the real system. Thus, one may be tempted to apply an informal validation. Concerning models of complex systems like socio-technical systems, this was examined in [41, 129]. Unfortunately, the quality of expert opinions has frequent deficits [40, 41]. The observed pronounced bias in the opinion of experts makes an informal validation of cognitive social engineering models questionable. Thus, formal validation methods are favored here in accordance with [42]. A full-scale formal validation of the overall model by analyzing the stochastic correspondence of the input-output behavior of model and real system seems to be doomed to failure as well, however.

Characterizing the behavior of cognitive systems usually requires a large number of parameters. Typically, the number of complete sets of observation data of the real system will be small compared to the number of data sets required for a comparison with adequate statistical significance. Moreover, the unique circumstances of each such observation hamper the comparability of reference data across different contexts.

We have to focus on less ambitious forms of a formal validation instead. One such method is to increase confidence to the model [130] by many unsuccessful tests to prove that it is wrong [131]. Informal lists of such tests can be found in [131, 132]. Due to the difficulties involved in the validation check of the overall behavior, it is recommended to put an emphasis on the microscopic scale. A system dynamics model consists of stocks and flows between the stocks. The flows describe the changes of the stocks and thus of the system state [133]. In this way, the flows define the dynamics of the model and thus the model behavior. Hence, a convenient validation strategy is a comparison between the flows in the model and the relationships observed in reality. As a complementary measure, one may look at selected model properties and compare them with the real system. Reproducing observed effects in the model and/or demonstrating the inability to generate unobserved effects are strong arguments for the validity of the model.

Supplementary to validating the simulation model, it is recommended to provide some insight into the risk calculation based on the model. The plain truth is that both humans and technical systems are subject to faults and failures. This leads naturally to the question to what extent the risk assessment process is erroneous as well. The quality of the result can be assured by several safeguards, however. Methods for supporting the generation of a complete (i.e., containing the most important failure modes) and correct simulation model have already been discussed above. Known errors, which cannot be removed due to fundamental reasons, have to be quantified. In the case of sampling errors, this is discussed in section 6.3. The provision of 'insight' into the calculation process is more challenging. Insight seems to be essential, however, because the risk assessment procedure will give just a single value (the statistical mean) or, at best, a value statistics. The proposed procedure does not provide, say, a list of effects contributing to this result in a form accessible by a human reader; providing just a lengthy list of system evolutions taken into account in the risk calculation does not enable a true human understanding of the risk calculation process. Up to the knowledge of the author, the development of methods supporting an understanding *how* the risk value results is still an open question. This challenge corresponds to the so-called transparency problem of AI [134–136].

8. SEMI-AUTONOMOUS CARS AS AN EXAMPLE OF APPLICATION

For showing the utility of the presented modeling concept, it is applied to modern semi-autonomous cars with in-car-entertainment. In this context, the term 'semi-autonomous' means that car systems *support* the primary driving task actively, but they do not *replace* the human driver. He still exerts an important control function. The investigation of the risk of driving such a car is usually confined to situations where due to technical faults the control of the vehicle is dispossessed from the driver at least partially. This perspective neglects the contribution of e.g., the interaction of the human driver with secondary systems like in-car-entertainment to the risk. Since the in-car-entertainment system is controlled by the driver as well, the driver's cognitive capabilities have to deal with driving task and entertainment system operation concurrently. Accordingly, effects such as decreased attention to or distraction from the driving task [137, 138] should be included in the risk assessment. The importance of this aspect seems to increase with the implemented automation level. Up to now, risk assessments practice such extended considerations not quite systematically. Since the presented modeling concept contains mechanisms describing the dynamics of attention (see section 5.4), multitasking errors and their consequences can be represented in the model and assessed from the risk perspective. This may improve the quality of risk assessments.

The given situation is even more complex, however. The growing importance of software systems in the automotive domain leads to numerous cyber security threats [139–142]. Due to the cyber-physical nature of semi-autonomous (or, more general, smart) cars, combined considerations of safety and security have become common, see for example [143–146]. Especially the in-car-entertainment is considered as a security problem [147], because it can be used by an attacker to infect the software systems of the car with malware [148]. Indeed, malware was already successfully used for attacking smart cars [149]. In the case of connected cars [150–152] integrated in a network, the network is one possible source of malware infections [153].

When using the in-car-entertainment system, the human driver may be confronted with a social engineering attempt [154], in which an attacker tries to infect the on-board software with malware [155]. Social engineering attacks are viewed as a serious threat in the automotive domain [156, 157] and addressed in several articles, e.g., in [158, 159]. They are discussed in more detail in [160–162]. Hidden technical connections between in-car-entertainment system and primary driving functionalities may exist and will enable a propagation of malware to safety-critical components. For example, the audio volume of the entertainment system sometimes has been coupled to an engine power sensor for assuring that the music sound drowns out the engine noise [163, p. 73].

When the driver becomes a victim of the social engineering attempt, the inadvertently installed malware may lead to additional risk items, which can be derived from the situation model and which are relevant in practice. We present three aspects from the perspective of cognition; for a more technically oriented discussion, the reader may be referred to e.g., [164, 165].

First, an entertainment system malfunctioning due to the malware may distract the driver from correctly perceiving the traffic and driving situation, because he may focus his attention on operating this system [166]. Thus, he is maybe not informed in a timely manner about a critical situation. Consequently, he may not react adequately. Such failure modes are the result of human weaknesses concerning the cognitive management of multitasking. In our modeling concept, corresponding aspects are represented in the subconsciousness part of the cognitive component M_C.

Second, an eventual malware infection can increase the criticality of the consequences of safety-related failures. In our modeling concept, such failures are in the realm of the technical system M_T. An example is a traffic or driving situation, which cannot be dealt with in the semi-autonomous mode. Then, the control of the vehicle has to be handed over from the artificial intelligence (AI) of the car to the human driver [167]. One option for indicating the pending control-handover is the usage of the audio system as a signaling device [168, 169]. If a malware has affected the functionality of the audio system, it may be impossible to trigger the corresponding warning sound. Consequently, the human driver may react too late or not at all resulting in an increase of the risk.

Third and finally, the malware may propagate to safety-critical components associated with the primary driving task and influence their function [170]. At least in the near future, the human driver will serve as principal fallback in such a case. For the currently practiced levels of automation, he must perform a continuous monitoring of the automated driving subtasks. If the driver detects an unacceptable driving behavior, he will switch off the semi-autonomous mode and take over himself. But what is unacceptable? We are talking here not necessarily about totally erratically driving actions. The technical component may just behave a little bit strange or uncommon from the viewpoint of the human driver. Is it a realistic assumption that an average human driver without any knowledge about computer science will recognize the malware attack and react accordingly? Some drivers will not have any insight into technical functions at all, which would profoundly aggravate the problem. Furthermore, after some time of practicing the autonomous mode, the human driver may perhaps overestimate its capabilities due to his trust in the AI. Accordingly, his average situation awareness may more and more decline with increasing time of usage [171, 172].

The above list of risk contributions indicates potentially significant corrections to the overall risk value, which may also be of interest from the theoretical point of view. Further elaboration seems to be important. This would require a detailed refinement of the presented modeling concept adapted to the automotive domain, followed by construction, implementation, and analysis of a model. The corresponding effort is beyond the scope of this chapter.

Summing up, the risk calculated on a purely technical basis and including only primary technical components may not correspond to the actual risk. The inclusion of interactions of the driver with vehicle [173] and human-related systems like in-car-entertainment [174] seems to be necessary. Cognitive-technical models may provide the corresponding corrections to the risk value. They may help to judge the numerous intricacies involved. For example, it seems to be a valid assumption that multimedia and web access in the car are used more often for higher automation levels. This may also increase the frequency of occurring security problems. Furthermore, an increased autonomy will also mean a larger attack surface for security threats. Both effects are opposed to the common assumption that higher automation will make cars safer. This contradiction deserves further analysis.

Conclusion

We have discussed how a social engineering situation can be assessed from the viewpoints of safety and security in a psychologically well-founded way. A modeling concept is presented, which includes both cognitive and technical aspects. The concept allows fast simulations, which enables Monte Carlo simulation-based risk assessments. The presented approach can take uncertainties of cognitive parameters into account; such uncertainties may be the result of e.g., the diversity of human personality. Hence, some progress seems to be made towards the inclusion of cognitive aspects in safety and security risk assessments. Since some details of cognitive psychology are still unclear, parts of the presented examination may be considered as preliminary. Their clarification will not only require further research in the psychological domain and the provision of quantitative risk values by an implemented model, but also a much larger and more detailed set of corresponding real-life data.

Due to the extent of the topic, it was necessary to focus on the most important aspects of social engineering and cognitive psychology. Several essential concerns have been excluded. They may be the topic of future research.

The author has barely touched the topic of social aspects. The human operator is a component, which is not belonging *exclusively* to the system under consideration. Outside working time, he becomes a part of other systems. He then belongs to the social network of his family, will have personal contacts on the way to work and so on. A bad dispute off the job may influence his performance in the job. His position in an organizational network or the safety and security culture practised at the workplace will have an influence as well [175]. If these aspects are considered as significant, the model has to be supplemented accordingly.

Since the discussion has been restricted to singular social engineering attempts, the capability of humans to learn and to adapt to new situations has been excluded. The model has to be extended, as soon as a series of social engineering attempts is considered. In effect, this will lead to a long-term dynamics of the cognitive characteristics.

Finally, the presented modeling concept focuses on human *errors*. This perspective neglects that humans are in the loop due to the superior quality of their higher cognitive functions like reasoning, knowledge processing, and problem solving capabilities. Unfortunately, these functions are not representable in a detailed way in a system dynamics model due to its sub-symbolic nature. This shortcoming may require a reconsideration of using a full-scale rule-based cognitive architecture as (part of) a cognitive model, which operates at the symbolic level. Indeed, the capabilities of such cognitive architectures have reached already a remarkable level. ACT-R, for example, could solve the Tower of Hanoi problem [69]. At the downside, however, one will pay for the improved capabilities of the cognitive model with a significant increase of the simulation runtime. Whether a risk assessment based on Monte Carlo simulation sampling will be still computationally tractable or not under these circumstances, is an open question.

ACKNOWLEDGMENTS

I would like to express my deep gratitude to Stephanie Öttl for her valuable and constructive suggestions. Additionally, the author is grateful for helpful remarks provided by Sabrina Dinkel, Furkan Eke, Angelika Froidl, Lukas Höhndorf, Martin Kaiser, Christoph Michalke, Julia Schaffer, Christine Schwarz-Hemmert, and Jessica Sturn. I want to thank Bastian Bernhardt for his support of creating the figures.

REFERENCES

[1] Christopher Hadnagy. *Social engineering: The art of human hacking*. John Wiley, 2010.

[2] Frank L Greitzer, Jeremy R Strozer, Sholom Cohen, Andrew P Moore, David Mundie, and Jennifer Cowley. Analysis of unintentional insider

threats deriving from social engineering exploits. In *Proceedings of Security and Privacy Workshops*, pages 236–250. IEEE, 2014.

[3] Kevin Orrey. A survey of USB exploit mechanisms, profiling stuxnet and the possible adaptive measures that could have made it more effective, 2011. online http://www.vulnerabilityassessment.co.uk/education/whitepaper.pdf.

[4] Steve Stasiukonis. Social engineering, the USB way. *Dark Reading*, 7, 2006. online http://www.darkreading.com/security/perimeter-security/208803634/index.html.

[5] Wenjun Fan, Lwakatare Kevin, and Rong Rong. Social engineering: I-E based model of human weakness for attack and defense investigations. *Computer Network and Information Security*, 9(1):1–11, 2017.

[6] Matthew Spinapolice. Mitigating the risk of social engineering attacks. 2011.

[7] Marcus Butavicius, Kathryn Parsons, Malcolm Pattinson, and Agata McCormac. Breaching the human firewall: Social engineering in phishing and spear-phishing emails. *preprint arXiv:1606.00887*, 2016.

[8] Robert Lee, Michael Assante, and Tim Connway. ICS CP/PE (cyber-to-physical or process effects) case study paper — German steel mill cyber attack. Technical report, 2014.

[9] Paul Mueller and Babak Yadegari. The stuxnet worm. Technical report, 2012.

[10] Sacha Brostoff and M Angela Sasse. Safe and sound: a safety-critical approach to security. In *Proceedings of the Workshop on New Security Paradigms*, pages 41–50. ACM, 2001.

[11] Jason Perno and Christian W Probst. Behavioural profiling in cyber-social systems. In *International Conference on Human Aspects of Information Security, Privacy, and Trust*, pages 507–517. Springer, 2017.

[12] Jason Smith, Selwyn Russell, and Mark Looi. Security as a safety issue in rail communications. In *Proceedings of the 8th Australian Workshop on*

Safety critical systems and software, pages 79–88. Australian Computer Society, 2003.

[13] Nancy Leveson. *Engineering a safer world: Systems thinking applied to safety*. MIT press, 2011.

[14] David Peter Eames and Jonathan Moffett. The integration of safety and security requirements. In *International Conference on Computer Safety, Reliability, and Security*, pages 468–480. Springer, 1999.

[15] Jean-Claude Laprie. Dependable computing: Concepts, limits, challenges. In *Special Issue of the 25th International Symposium On Fault-Tolerant Computing*, pages 42–54, 1995.

[16] Yulia Cherdantseva, Pete Burnap, Andrew Blyth, Peter Eden, Kevin Jones, Hugh Soulsby, and Kristan Stoddart. A review of cyber security risk assessment methods for SCADA systems. *Computers & security*, 56:1–27, 2016.

[17] Ian Mann. *Hacking the human: social engineering techniques and security countermeasures*. Routledge, 2017.

[18] David FC Brewer. Applying security techniques to achieving safety. In *Directions in Safety-Critical Systems*, pages 246–256. Springer, 1993.

[19] Erland Jonsson and Tomas Olovsson. On the integration of security and dependability in computer systems. In *Proceedings of the IASTED International Conference on Reliability, Quality Control and Risk Assessment Washington DC, USA*, pages 93–97, 1992.

[20] Karin Sallhammar, Bjarne E Helvik, and Svein J Knapskog. On stochastic modeling for integrated security and dependability evaluation. *JNW*, 1(5):31–42, 2006.

[21] Enrico Zio. The future of risk assessment. *Reliability Engineering & System Safety*, 177:176–190, 2018.

[22] Kristina Yordanova. Toward a unified human behaviour modelling approach. Technical report, Institut für Informatik, Universität Rostock, 2011. CS-02-11.

[23] Kenny Jansson. *A model for cultivating resistance to social engineering attacks*. PhD thesis, Nelson Mandela Metropolitan University, 2011.

[24] CERT Insider Threat Team. Unintentional insider threats: Social engineering. Technical report, Software Engineering Institute, Carnegie Mellon University, 2014. CMU/SEI-2013-TN-024.

[25] Koteswara Ivaturi. *Social engineering — emerging attacks, awareness and impact on online user attitudes and behaviours*. PhD thesis, ResearchSpace@Auckland, 2014.

[26] Peter M Bednar and Vasilios Katos. Addressing the human factor in information systems security. In *MCIS*, page 72, 2009.

[27] Marcus Nohlberg and Stewart Kowalski. The cycle of deception: a model of social engineering attacks, defenses and victims. In *Proccedings of the Second International Symposium on Human Aspects of Information Security and Assurance*, pages 1–11. University of Plymouth, 2008.

[28] Marcus Nohlberg. *Securing information assets: understanding, measuring and protecting against social engineering attacks*. PhD thesis, KTH, 2008.

[29] Vidar Engmo. Representation of human behavior in military simulations. Master's thesis, Institute for telematic, 2008.

[30] Daniel V Holt and Magda Osman. Approaches to cognitive modeling in dynamic systems control. *Frontiers in psychology*, 8:2032, 2017.

[31] KR Laughery, Christian Lebiere, and Susan Archer. Modeling human performance in complex systems. *Handbook of human factors and ergonomics*, pages 965–996, 2006.

[32] Donald Abrahamson and Adrian L Sepeda. Expanding known process safety and risk analysis concepts to manage security concerns. In *Proceedings of the 40th Loss prevention Symposium*, 2006.

[33] C Warren Axelrod. *Engineering safe and secure software systems*. Artech House, 2012.

[34] Maria B Line, Odd Nordland, Lillian Røstad, and Inger Anne Tøndel. Safety vs security? In *PSAM conference, New Orleans, USA*, 2006.

[35] Ludovic Pietre-Cambacedes and Claude Chaudet. Disentangling the relations between safety and security. In *Proceedings of the 9th international conference on Applied informatics and communications*, pages 156–161. World Scientific and Engineering Academy and Society (WSEAS), 2009.

[36] Terje Aven. A unified framework for risk and vulnerability analysis covering both safety and security. *Reliability engineering & System safety*, 92(6):745–754, 2007.

[37] Øystein Amundrud, Terje Aven, and Roger Flage. How the definition of security risk can be made compatible with safety definitions. *Proceedings of the Institution of Mechanical Engineers, Part O: Journal of Risk and Reliability*, 231(3):286–294, 2017.

[38] Ludovic Piètre-Cambacédès and Marc Bouissou. Cross-fertilization between safety and security engineering. *Reliability Engineering & System Safety*, 110:110–126, 2013.

[39] Koen Buyens, Bart De Win, and Wouter Joosen. Empirical and statistical analysis of risk analysis-driven techniques for threat management. In *Proceedings of the Second International Conference on Availability, Reliability and Security (ARES)*, pages 1034–1041. IEEE, 2007.

[40] Simon R Goerger. Validating computational human behavior models: consistency and accuracy issues. Technical report, Naval postgraduate school, Monterey CA, 2004.

[41] Scott Y Harmon, CWD Hoffman, Avelino J Gonzalez, Rainer Knauf, and Valerie B Barr. Validation of human behavior representations. In *Proceedings of the Workshop Foundations for V&V in the 21st Century*, pages 22–4, 2002.

[42] Michael Cebulla. Using advanced formal concepts in interdisciplinary analysis and design of safety-critical sociotechnical systems. In *Proceedings of the 26th Annual Computer Software and Applications Conference*, 2002.

[43] Ludovic Apvrille and Yves Roudier. Towards the model-driven engineering of secure yet safe embedded systems. In *Proceedings of the First International Workshop on Graphical Models for Security*, 2014.

[44] André Arnold, Gérald Point, Alain Griffault, and Antoine Rauzy. The altarica formalism for describing concurrent systems. *Fundamenta Informaticae*, 40(2, 3):109–124, 1999.

[45] Siwar Kriaa, Ludovic Pietre-Cambacedes, Marc Bouissou, and Yoran Halgand. A survey of approaches combining safety and security for industrial control systems. *Reliability Engineering & System Safety*, 139:156–178, 2015.

[46] J Devooght. Dynamic reliability: the challenges ahead. In *Fifth International Workshop on Dynamic Reliability: Future Directions*, 1998.

[47] Enrico Zio. Reliability engineering: Old problems and new challenges. *Reliability Engineering & System Safety*, 94(2):125–141, 2009.

[48] Hans De Bruijn and Paulien M Herder. System and actor perspectives on sociotechnical systems. *IEEE Transactions on systems, man, and cybernetics-part A: Systems and Humans*, 39(5):981–992, 2009.

[49] Alexis Morris. Socio-technical systems in ICT: a comprehensive survey. Technical report, University of Trento, 2009.

[50] James Ladyman, James Lambert, and Karoline Wiesner. What is a complex system? *European Journal for Philosophy of Science*, 3(1):33–67, 2013.

[51] John Sterman. System dynamics: systems thinking and modeling for a complex world. 2002.

[52] Jay W Forrester. System dynamics, systems thinking, and soft OR. *System dynamics review*, 10(2-3):245–256, 1994.

[53] Patrick Einzinger. *A Comparative Analysis of System Dynamics and Agent-Based Modelling for Health Care Reimbursement Systems*. PhD thesis, 2014.

[54] Rudolph Oosthuizen, Malinkeng M Molekoa, and Francois Mouton. System dynamics modelling to investigate the cost-benefit of cyber security investment. 2018.

[55] Donella H Meadows, Donella H Meadows, Jørgen Randers, and William W Behrens III. The limits to growth: a report to the club of Rome. *accessed via Google Scholar*, 1972.

[56] Uma Kannan. *Cyber Security System Dynamic Modeling*. PhD thesis, Auburn University, 2017.

[57] Joachim Draeger and Stephanie Öttl. Malware epidemics effects in a lanchester conflict model. *preprint arXiv:1811.01892*, 2018.

[58] Farhad Foroughi. The application of system dynamics for managing information security insider-threats of IT organization. In *Proceedings of the World Congress on Engineering*, volume 1, pages 2–4. Citeseer, 2008.

[59] Carlos Melara, Jose Maria Sarriegui, Jose J Gonzalez, Agata Sawicka, and David L Cooke. A system dynamics model of an insider attack on an information system. In *Proceedings of the 21st International Conference of the System dynamics Society*, pages 20–24, 2003.

[60] Sang-Chin Yang and Yi-Lu Wang. System dynamics based insider threats modeling. *International Journal of Network Security and its Applications*, 3(3):1–14, 2011.

[61] Mehrnaz Akbari Roumani, Chun Che Fung, and Pema Choejey. Assessing economic impact due to cyber attacks with system dynamics approach. In *Proceedings of the 12th International Conference on Electrical Engineering/Electronics, Computer, Telecommunications and Information Technology (ECTI-CON)*, pages 1–6. IEEE, 2015.

[62] Keon Hee Lee. *Development of a System Dynamics Model for the Assessment of Nuclear Security Culture at a Nuclear Power Plant*. PhD thesis, Seoul National University, 2016.

[63] Andrew P Moore, Dawn M Cappelli, Hannah Joseph, and Randall F Trzeciak. An experience using system dynamics to facilitate an insider threat workshop. In *Proceedings of the 25th International Conference of the System Dynamics Society*, 2007.

[64] Andrew P Moore, David A Mundie, and Matthew L Collins. A system dynamics model for investigating early detection of insider threat risk.

In *Conference Proceedings of the 31st International Conference of the System Dynamics Society. Cambridge, MA*, pages 978–1, 2013.

[65] Zhenhua Cai, Ben Goertzel, Changle Zhou, Yongfeng Zhang, Min Jiang, and Gino Yu. Dynamics of a computational affective model inspired by Dörner's psi theory. *Cognitive Systems Research*, 17:63–80, 2012.

[66] Allen Newell. *Unified theories of cognition*. Harvard University Press, 1994.

[67] Falk Lieder and Thomas L Griffiths. Resource-rational analysis: understanding human cognition as the optimal use of limited computational resources. *Psychological Review*, 85(4):249–277, 1991.

[68] John E Morrison. A review of computer-based human behavior representations and their relation to military simulations. Technical report, Institute for Defense Analysis, Alexandria VA, 2003.

[69] Hui-Qing Chong, Ah-Hwee Tan, and Gee-Wah Ng. Integrated cognitive architectures: a survey. *Artificial Intelligence Review*, 28(2):103–130, 2007.

[70] Frank E Ritter, Nigel R Shadbolt, David Elliman, Richard M Young, Fernand Gobet, and Gordon D Baxter. Techniques for modeling human performance in synthetic environments: A supplementary review. Technical report, Human systems Information analysis center, Wright-Patterson AFB, 2003.

[71] Wouter Lotens, Laurel Allender, Joe Armstrong, Andrew Belyavin, Brad Cain, Martin Castor, Kevin Gluck, Wolf KÃppler, Peter Kwantes, Mikael Lundin, Gina Thomas, and Niklas Wallin. Human behavior representation in constructive simulation. Technical report, NATO, 2009. TR-HFM-128.

[72] Kenneth Leiden, K Ronald Laughery, John Keller, Jon French, Walter Warwick, and Scott D Wood. A review of human performance models for the prediction of human error. *Ann Arbor*, 1001:48105, 2001.

[73] Jeffrey M Stanton, Kathryn R Stam, Paul Mastrangelo, and Jeffrey Jolton. Analysis of end user security behaviors. *Computers & security*, 24(2):124–133, 2005.

[74] Detmar W Straub and Richard J Welke. Coping with systems risk: security planning models for management decision making. *MIS quarterly*, pages 441–469, 1998.

[75] Katrina M Groth and Ali Mosleh. A data-informed PIF hierarchy for model-based human reliability analysis. *Reliability Engineering & System Safety*, 108:154–174, 2012.

[76] Frank E Ritter and MN Avramides. Steps towards including behavior moderators in human performance models in synthetic environments. *Penn State*, 2000.

[77] Robert Feyen and Yili Liu. Modeling task performance using the queuing network-model human processor (QN-MHP). In *Proceedings of the 4th International Conference on Cognitive Modeling*, pages 73–78, 2001.

[78] Yili Liu, Robert Feyen, and Omer Tsimhoni. Queueing network-model human processor (QN-MHP): A computational architecture for multitask performance in human-machine systems. *Ann Arbor*, 1001:48109, 2004.

[79] Yili Liu, Robert Feyen, and Omer Tsimhoni. Queueing network-model human processor (QN-MHP): A computational architecture for multitask performance in human-machine systems. *ACM Transactions on Computer-Human Interaction (TOCHI)*, 13(1):37–70, 2006.

[80] Omer Tsimhoni and Yili Liu. Modeling steering using the queueing network-model human processor (QN-MHP). In *Proceedings of the human factors and ergonomics society annual meeting*, volume 47, pages 1875–1879. SAGE Publications, 2003.

[81] Steven A Sloman. The empirical case for two systems of reasoning. *Psychological Bulletin*, 119(1):3, 1996.

[82] Jonathan St BT Evans. Dual-processing accounts of reasoning, judgment, and social cognition. *Annu. Rev. Psychol.*, 59:255–278, 2008.

[83] Martin Hilbert. Toward a synthesis of cognitive biases: How noisy information processing can bias human decision making. *Psychological Bulletin*, 138(2):211, 2012.

[84] MP Fewell and Mark G Hazen. Cognitive issues in modelling network-centric command and control. Technical Report DSTO-RR-0293, DSTO, 2005.

[85] Peter M Todd and Gerd Gigerenzer. Précis of simple heuristics that make us smart. *Behavioral and brain sciences*, 23(5):727–741, 2000.

[86] Amos Tversky and Daniel Kahneman. Judgment under uncertainty: Heuristics and biases. *science*, 185(4157):1124–1131, 1974.

[87] Douglas P Twitchell. Social engineering and its countermeasures. In *Handbook of research on social and organizational liabilities in information security*, pages 228–242. IGI Global, 2009.

[88] Daniel Kahneman. *Attention and effort*. Prentice Hall, 1973.

[89] Christopher D Wickens. Processing resources and attention. *Multiple-task performance*, pages 3–34, 1991.

[90] Christopher D Wickens. Multiple resources and mental workload. *Human factors*, 50(3):449–455, 2008.

[91] Richard C Atkinson and Richard M Shiffrin. Human memory: A proposed system and its control processes. In *Psychology of learning and motivation*, volume 2, pages 89–195. Elsevier, 1968.

[92] Chris Kimble. Knowledge management, codification and tacit knowledge. *Information Research*, 18(2), 2013.

[93] Thomas L Griffiths, Falk Lieder, and Noah D Goodman. Rational use of cognitive resources: Levels of analysis between the computational and the algorithmic. *Topics in cognitive science*, 7(2):217–229, 2015.

[94] Bernd Schmidt and Bernhard Schneider. Agent-based modelling of human acting, deciding and behaviour-the reference model PECS. In *Proceedings 18th European Simulation Multiconference*, 2004.

[95] Yingxu Wang and Vincent Chiew. On the cognitive process of human problem solving. *Cognitive systems research*, 11(1):81–92, 2010.

[96] Yingxu Wang, Shushma Patel, Dilip Patel, and Ying Wang. A layered reference model of the brain. In *Proceedings of the 2nd International Conference on Cognitive Informatics*, pages 7–17. IEEE, 2003.

[97] Yingxu Wang, Ying Wang, Shushma Patel, and Dilip Patel. A layered reference model of the brain (LRMB). *IEEE Transactions on Systems, Man, and Cybernetics, Part C (Applications and Reviews)*, 36(2):124–133, 2006.

[98] Christoph Urban. *Das Referenzmodell PECS: Agentenbasierte Modellierung menschlichen Handelns, Entscheidens und Verhaltens [The reference model PECS: Agent-based modeling of human action, decision, and behavior]*. PhD thesis, University of Passau, 2005.

[99] Harald Schaub. Human personality as an information processing system. In *System Dynamics Conference*. System Dynamics Society, 2003.

[100] Harald Schaub. *Persönlichkeit und Problemlösen [Personality and problem solving]*. PhD thesis, University of Bamberg, 1998.

[101] Alexei Sharpanskykh. Integrated modeling of cognitive agents in socio-technical systems. In *Proceedings of the International Symposium on Agent and Multi-Agent Systems: Technologies and Applications*, pages 262–271. Springer, 2010.

[102] Abdullah Ayed M Algarni. *The impact of source characteristics on users' susceptibility to social engineering victimization in social networks*. PhD thesis, Queensland University of Technology, 2016.

[103] Santos M Galvez and Indira R Guzman. Identifying factors that influence corporate information security behavior. *Proceedings of AMCIS*, 2009. Paper 765.

[104] Dietrich Dörner. Eine Systemtheorie der Motivation [A system theory of motivation]. In Julius Kuhl and Heinz Heckhausen, editors, *Motivation, Volition und Handlung*, pages 329–356. Hogrefe, 1994.

[105] Ye Diana Wang and Henry H Emurian. An overview of online trust: Concepts, elements, and implications. *Computers in human behavior*, 21(1):105–125, 2005.

[106] K Jansson and R Von Solms. Social engineering: towards a holistic solution. In *Proceedings of the South African Information Security Multi-Conference: Port Elizabeth, South Africa, 17-18 May 2010*, page 23. Lulu.com, 2011.

[107] Daniel Scherer, Q Vieira Maria de Fátima, and José Alves do N Neto. Human error categorization: An extension to classical proposals applied to electrical systems operations. In *IFIP Human-Computer Interaction Symposium*, pages 234–245. Springer, 2010.

[108] Stanley Kaplan and B John Garrick. On the quantitative definition of risk. *Risk analysis*, 1(1):11–27, 1981.

[109] Joachim Draeger. Roadmap to a unified treatment of safety and security. In *Proceedings of the 10th Conference on System Safety and Cyber Security*. IET, 2015.

[110] Joachim Draeger. Formalized risk assessment for safety and security. *preprint arXiv:1709.00567*, 2017.

[111] Charlie C Chen, RS Shaw, and Samuel C Yang. Mitigating information security risks by increasing user security awareness: A case study of an information security awareness system. *Information Technology, Learning & Performance Journal*, 24(1), 2006.

[112] Tzipora Halevi, Jim Lewis, and Nasir Memon. Phishing, personality traits and facebook. *preprint arXiv:1301.7643*, 2013.

[113] Michael Workman. Wisecrackers: A theory-grounded investigation of phishing and pretext social engineering threats to information security. *Journal of the American Society for Information Science and Technology*, 59(4):662–674, 2008.

[114] Vanderley de Vasconcelos, Wellington Antonio Soares, Antônio Carlos Lopes da Costa, and Amanda Laureano Raso. Treatment of uncertainties in probabilistic risk assessment. In *System Reliability*. IntechOpen, 2019.

[115] Robert L Winkler. Uncertainty in probabilistic risk assessment. *Reliability Engineering & System Safety*, 54(2-3):127–132, 1996.

[116] Terje Aven and Enrico Zio. Some considerations on the treatment of uncertainties in risk assessment for practical decision making. *Reliability Engineering & System Safety*, 96(1):64–74, 2011.

[117] Terje Aven. Trends in quantitative risk assessments. *International Journal of Performability Engineering*, 5(5), 2009.

[118] Terje Aven and Bodil S Krohn. A new perspective on how to understand, assess and manage risk and the unforeseen. *Reliability Engineering & System Safety*, 121:1–10, 2014.

[119] Fabrizio Baiardi, Fabio Corò, Federico Tonelli, Alessandro Bertolini, Roberto Bertolotti, and Daniela Pestonesi. Assessing and managing ICT risk with partial information. In *Proceedings of 16th IEEE International Conference on High Performance Computing and Communications, HPCC, 6th IEEE International Symposium on Cyberspace Safety and Security, CSS, and 11th IEEE International Conference on Embedded Software and Systems, ICESS*, pages 1213–1220. IEEE, 2014.

[120] Fabrizio Baiardi, Fabio Corò, Federico Tonelli, and Daniele Sgandurra. A scenario method to automatically assess ICT risk. In *Proceedings of the 22nd Euromicro International Conference on Parallel, Distributed, and Network-Based Processing*, pages 544–551. IEEE, 2014.

[121] Fabrizio Baiardi, Federico Tonelli, and AD Ruggiero Di Biase. Assessing and managing risk by simulating attack chains. In *Proceedings of the 24th Euromicro International Conference on Parallel, Distributed, and Network-Based Processing (PDP)*, pages 581–584. IEEE, 2016.

[122] Marzio Marseguerra and Enrico Zio. Basics of the Monte Carlo method with application to system reliability. 2002.

[123] Marzio Marseguerra and Enrico Zio. Monte Carlo approach to PSA for dynamic process systems. *Reliability Engineering & System Safety*, 52(3):227–241, 1996.

[124] Alejandro Llorente Pinto et al. Analysis of the convergence of Monte Carlo averages. Master's thesis, 2012.

[125] Ray Waller. *Low-Probability High-Consequence Risk Analysis: Issues, Methods, and Case Studies*, volume 2. Springer Science & Business Media, 2013.

[126] Bernard Lapeyre. Introduction to Monte-Carlo methods. *Lecture, Halmstad, Sweden*, pages 2–4, 2007.

[127] Robert G Sargent. Verification and validation of simulation models. In *Proceedings of the 2010 Winter Simulation Conference*, pages 166–183. IEEE, 2010.

[128] Gunnar Johannsen, Alexander H Levis, and Henk G Stassen. Theoretical problems in man-machine systems and their experimental validation. *Automatica*, 30(2):217–231, 1994.

[129] Avelino J Gonzalez and Maureen Murillo. Validation of human behavioral models. In *Proceedings of FLAIRS*, pages 489–493, 1999.

[130] Charles Featherston, Matthew Doolan, et al. A critical review of the criticisms of system dynamics. 2012.

[131] JD Sterman. Business dynamics: systems thinking and modeling for a complex world: Jeffrey j. *Shelstad*, 2000.

[132] David Walter Peterson. *Hypothesis, estimation, and validation of dynamic social models: energy demand modeling*. PhD thesis, Massachusetts Institute of Technology, 1975.

[133] Uma Kannan, Rajendran Swamidurai, and David Umphress. System dynamics as a tool for modeling application layer cyber security. In *Proceedings of the International Conference on Modeling, Simulation and Visualization Methods (MSV)*, page 88, 2016.

[134] Sandra Wachter, Brent Mittelstadt, and Luciano Floridi. Transparent, explainable, and accountable AI for robotics. *Science Robotics*, 2(6), 2017.

[135] Robert H Wortham, Andreas Theodorou, and Joanna J Bryson. What does the robot think? transparency as a fundamental design requirement for intelligent systems. In *IJCAI — ethics for artificial intelligence workshop*, 2016.

[136] Keng Siau and Weiyu Wang. Building trust in artificial intelligence, machine learning, and robotics. *Cutter Business Technology Journal*, 31(2):47–53, 2018.

[137] Giovanna Broccia. Model-based analysis of driver distraction by infotainment systems in automotive domain. In *Proceedings of the SIGCHI Symposium on Engineering Interactive Computing Systems*, pages 133–136. ACM, 2017.

[138] Michael Geiger, Martin Zobl, Klaus Bengler, and Manfred Lang. Intermodal differences in distraction effects while controlling automotive user interfaces. In *Proc. Int. Conf. on Human-Computer Interaction HCI, New Orleans, Louisiana, USA*, 2001.

[139] Abdulmalik Humayed and Bo Luo. Cyber-physical security for smart cars: taxonomy of vulnerabilities, threats, and attacks. In *Proceedings of the Sixth International Conference on Cyber-Physical Systems*, pages 252–253. ACM, 2015.

[140] Michael Huber, Michael Brunner, Clemens Sauerwein, Carmen Carlan, and Ruth Breu. Roadblocks on the highway to secure cars: An exploratory survey on the current safety and security practice of the automotive industry. In *International Conference on Computer Safety, Reliability, and Security*, pages 157–171. Springer, 2018.

[141] KV Harish and B Amutha. Survey on security in autonomous cars. In *International Conference on Communications and Cyber Physical Engineering 2018*, pages 629–637. Springer, 2018.

[142] Ivan Studnia, Vincent Nicomette, Eric Alata, Yves Deswarte, Mohamed Kaâniche, and Youssef Laarouchi. Survey on security threats and protection mechanisms in embedded automotive networks. In *43rd Annual Conference on Dependable Systems and Networks Workshop (DSN-W)*, pages 1–12. IEEE, 2013.

[143] Georg Macher, Andrea Höller, Harald Sporer, Eric Armengaud, and Christian Kreiner. A combined safety-hazards and security-threat analysis method for automotive systems. In *International Conference on Computer Safety, Reliability, and Security*, pages 237–250. Springer, 2014.

[144] Peter H Jesty and David D Ward. Towards a unified approach to safety and security in automotive systems. In *The Safety of Systems*, pages 21–34. Springer, 2007.

[145] Tiago Amorim, Helmut Martin, Zhendong Ma, Christoph Schmittner, Daniel Schneider, Georg Macher, Bernhard Winkler, Martin Krammer, and Christian Kreiner. Systematic pattern approach for safety and security co-engineering in the automotive domain. In *International Conference on Computer Safety, Reliability, and Security*, pages 329–342. Springer, 2017.

[146] Jürgen Dürrwang, Kristian Beckers, and Reiner Kriesten. A lightweight threat analysis approach intertwining safety and security for the automotive domain. In *International Conference on Computer Safety, Reliability, and Security*, pages 305–319. Springer, 2017.

[147] Pratik Satam, Jesus Pacheco, Salim Hariri, and Mohommad Horani. Autoinfotainment security development framework (ASDF) for smart cars. In *International Conference on Cloud and Autonomic Computing (IC-CAC)*, pages 153–159. IEEE, 2017.

[148] Joongyong Choi and Seong-il Jin. Security threats in connected car environment and proposal of in-vehicle infotainment-based access control mechanism. In *Advanced Multimedia and Ubiquitous Engineering*, pages 383–388. Springer, 2018.

[149] K Strandberg. Avoiding vulnerabilities in connected cars. Master's thesis, Chalmers University of Technology, 2016.

[150] Mushabbar Hussain. Security in connected cars. In *Proceedings of the European Automotive Congress EAEC-ESFA*, pages 267–275. Springer, 2016.

[151] Prabhat Ram, Jouni Markkula, Ville Friman, and Arian Raz. Security and privacy concerns in connected cars: A systematic mapping study. In *44th Euromicro Conference on Software Engineering and Advanced Applications (SEAA)*, pages 124–131. IEEE, 2018.

[152] Tim Ring. Connected cars — the next target for hackers. *Network Security*, 2015(11):11–16, 2015.

[153] Tao Zhang, Helder Antunes, and Siddhartha Aggarwal. Defending connected vehicles against malware: Challenges and a solution framework. *IEEE Internet of Things journal*, 1(1):10–21, 2014.

[154] Edwin Franco Myloth Josephlal and Sridhar Adepu. Vulnerability analysis of an automotive infotainment system's wifi capability. In *19th International Symposium on High Assurance Systems Engineering (HASE)*, pages 241–246. IEEE, 2019.

[155] Václav Linkov, Petr Zámečník, Darina Havlíčková, and Chih-Wei Pai. Human factors in the cybersecurity of autonomous cars: Trends in current research. *Frontiers in psychology*, 10:995, 2019.

[156] Stephen Checkoway, Damon McCoy, Brian Kantor, Danny Anderson, Hovav Shacham, Stefan Savage, Karl Koscher, Alexei Czeskis, Franziska Roesner, Tadayoshi Kohno, et al. Comprehensive experimental analyses of automotive attack surfaces. In *USENIX Security Symposium*, volume 4, pages 447–462. San Francisco, 2011.

[157] Alex Oyler and Hossein Saiedian. Security in automotive telematics: a survey of threats and risk mitigation strategies to counter the existing and emerging attack vectors. *Security and Communication Networks*, 9(17):4330–4340, 2016.

[158] Tobias Hoppe, Stefan Kiltz, and Jana Dittmann. Security threats to automotive can networks–practical examples and selected short-term countermeasures. In *International Conference on Computer Safety, Reliability, and Security*, pages 235–248. Springer, 2008.

[159] Marko Wolf. Attackers and attacks in the automotive domain. In *Security Engineering for Vehicular IT Systems*, pages 77–89. Springer, 2009.

[160] Arslan Munir and Farinaz Koushanfar. Design and analysis of secure and dependable automotive CPS: A steer-by-wire case study. *IEEE Transactions on Dependable and Secure Computing*, 2018.

[161] Remy Spaan, Lejla Batina, Peter Schwabe, and Sjoerd Verheijden. Secure updates in automotive systems. *Nijmegen: Radboud University*, pages 1–71, 2016.

[162] Bikash Poudel and Arslan Munir. Design and evaluation of a reconfigurable ECU architecture for secure and dependable automotive CPS. *IEEE Transactions on Dependable and Secure Computing*, 2018.

[163] Stefan Hamerich. *Sprachbedienung im Automobil: Teilautomatisierte Entwicklung benutzerfreundlicher Dialogsysteme [Voice-command operation in Cars: Semi-automated development of user-friendly dialogue systems]*. Springer-Verlag, 2010.

[164] Zhendong Ma, Walter Seböck, Bettina Pospisil, Christoph Schmittner, and Thomas Gruber. Security and privacy in the automotive domain: A technical and social analysis. In *International Conference on Computer Safety, Reliability, and Security*, pages 427–434. Springer, 2017.

[165] Roland E Haas and Dietmar PF Möller. Automotive connectivity, cyber attack scenarios and automotive cyber security. In *International Conference on Electro Information Technology (EIT)*, pages 635–639. IEEE, 2017.

[166] Marcel Reichersdoerfer. Driver distraction in automotive HMI. *Institute of Media Informatics Ulm University*, page 33, 2012.

[167] Chasel Lee. Grabbing the wheel early: Moving forward on cybersecurity and privacy protections for driverless cars. *Federal Communications Law Journal*, 69, 2017.

[168] Andreas Mueller, Markus Ogrizek, Lukas Bier, and Bettina Abendroth. Design concept for a visual, vibrotactile and acoustic take-over request in a conditional automated vehicle during non-driving-related tasks. In *6th international conference on driver distraction and inattention, Gothenburg, Sweden*, volume 2, 2018.

[169] Pavlo Bazilinskyy and Joost de Winter. Auditory interfaces in automated driving: an international survey. *PeerJ Computer Science*, 1:e13, 2015.

[170] Hirofumi Onishi. Cyber security for automotive electronics. In *ITS America 22nd Annual Meeting & Exposition*, 2012.

[171] Stephen M Casner, Edwin L Hutchins, and Don Norman. The challenges of partially automated driving. *Communications of the ACM*, 59(5):70–77, 2016.

[172] Simon Chesterman. Artificial intelligence and the problem of autonomy. *Notre Dame Journal on Emerging Technologies*, 1, 2019.

[173] Pierre Van Elslande, Claire L Naing, and Ralf Engel. Analyzing human factors in road accidents: Trace wp5 summary report. 2008.

[174] Bastian Pfleging, Maurice Rang, and Nora Broy. Investigating user needs for non-driving-related activities during automated driving. In *Proceedings of the 15th international conference on mobile and ubiquitous multimedia*, pages 91–99. ACM, 2016.

[175] E Frumento, R Puricelli, F Freschi, D Ariu, N Weiss, C Dambra, I Cotoi, P Roccetti, M Rodriguez, L Adrei, et al. The role of social engineering in evolution of attacks. Technical report, 2016.

INDEX

A

access to early years education, 82, 83, 84, 85, 107, 108
acquisition of knowledge, 87
AI/IoT, v, vii, viii, 1, 2, 3, 19, 32, 33, 65, 69
AI/IoT era, viii, 2, 3, 19, 32, 33, 65
artificial intelligence (AI), v, vii, viii, 1, 2, 3, 5, 19, 10, 20, 27, 28, 30, 32, 33, 35, 37, 61, 62, 64, 65, 66, 67, 68, 69, 74, 75, 76, 138, 140, 155, 156
assurance case, 2, 3, 13, 14, 29, 31, 32, 33, 34, 38, 39, 46, 47, 48, 62, 65, 67, 68, 71, 73, 74, 77, 78, 79
attention, 22, 114, 116, 117, 118, 121, 122, 125, 126, 127, 128, 129, 138, 139, 151
awareness, 88, 106, 108, 134, 140, 145, 153

B

background information, 123, 131
black market, 20

C

Causal Analysis using System Theory (CAST), viii, 1, 2, 3, 10, 33, 36, 39, 65, 70
child labour, 84, 112
cognition, 114, 123, 124, 139, 149, 150
cognitive biases, 125, 150
cognitive function, 142
cognitive models, 124
cognitive performance, 124
cognitive process, 117, 126, 129, 151
cognitive psychology, ix, 115, 116, 141
cognitive science, 151
cognitive system, 137
cognitive tasks, 117, 118, 125
collaboration, 102, 105
common criteria (CC), v, viii, 1, 2, 3, 14, 15, 16, 22, 28, 29, 30, 31, 32, 33, 34, 35, 36, 37, 38, 39, 40, 46, 47, 48, 49, 61, 62, 63, 65, 66, 67, 68, 69, 70, 71, 73, 75, 77, 78
corroborate, 91, 92
cybersecurity, 10, 61, 68, 158, 159

D

difficult circumstances, v, viii, 81, 82, 83, 85, 86
disadvantaged backgrounds, 83
dry areas, 83

E

early years education, viii, 81, 83, 84, 85, 92, 101, 107, 108
early years school, v, 81, 85, 86, 89, 92, 94, 95, 97, 103, 105
eavesdropping, 20, 21
effects of disasters, 84
ethnic conflicts, ix, 82, 96, 97, 105

F

Fault Tree Analysis, 9, 69
Functional Resonance Analysis Method (FRAM), vii, 1, 2, 3, 10, 12, 32, 33, 36, 61, 62, 63, 65, 67, 68, 70

G

Goal Structuring Notation (GSN), viii, 1, 2, 13, 14, 33, 34, 39, 40, 46, 47, 61, 62, 65, 66, 67, 68, 73
government support, 101, 102

H

harsh weather, ix, 82, 94, 97, 105
hazardous, 94
hostile social environment, 85

I

indigenous knowledge, 103, 104, 108
information processing, 152
intentionality, 124, 128
IoT, vii, 1, 2, 3, 16, 19, 21, 22, 23, 24, 25, 26, 27, 28, 29, 30, 32, 36, 39, 56, 61, 64, 65, 67, 68, 73, 77, 78

J

Japan, 1, 16, 20, 21, 24, 25, 27, 70, 75, 76, 77, 78

L

long distance, 83, 84, 90, 92, 93, 94, 97, 108
low participation rates, 85

M

malware, 57, 59, 114, 115, 116, 118, 120, 121, 122, 123, 135, 139, 140, 148, 158
marginalized areas, 83
Microsoft, 6, 42, 74
mobile schools, ix, 82, 98, 99, 100, 102, 103, 106, 107, 108
Monte Carlo method, 135, 154
Monte Carlo simulation, ix, 114, 117, 142

N

nomadic pastoralist, 84, 88, 90, 96, 100, 104, 105, 106, 107, 108

O

occupation, 87, 88, 99
outbreak of diseases, 84

Index

P

parents, v, vii, viii, 81, 82, 85, 86, 87, 88, 89, 90, 91, 92, 93, 94, 95, 96, 97, 98, 99, 101, 102, 103, 104, 105, 106, 108
pasture, 93, 96, 98
preventing, v, vii, viii, 36, 81, 82, 85, 86, 89, 92, 94, 95, 97, 98, 105
probability, ix, 20, 21, 114, 123, 126, 128, 129, 130, 132, 133, 134, 135, 136
probability distribution, ix, 114, 134
probability theory, 134
problem solving, 68, 151
procedural knowledge, 126
psychology, 113, 116, 119, 145, 158

Q

quality assurance, 28, 40, 68
quantitative technique, viii, 82

R

reliability, 15, 24, 25, 26, 27, 28, 32, 36, 68, 91, 147, 150, 154
resilience engineering, vii, 1, 2, 3, 10, 11, 36, 70
risk assessment, v, ix, 113, 114, 115, 116, 117, 118, 119, 121, 124, 126, 130, 134, 135, 136, 138, 139, 141, 142, 144, 153, 154
risk management, 10, 11
robotics, 155, 156
rugged terrain, ix, 82, 95, 97, 105, 106

S

safety and security challenges, vii, viii, 82, 86, 97
saline water, 96, 97
scenario function, viii, 1, 2, 3, 16, 17, 32, 33, 37, 39, 63, 67, 68
school dropout, 83, 85
secondary education, 110
secondary school students, 110
security, vii, viii, ix, 1, 2, 3, 6, 7, 8, 10, 13, 14, 15, 16, 19, 20, 21, 22, 23, 24, 26, 28, 29, 30, 31, 32, 33, 34, 36, 37, 39, 40, 41, 43, 46, 47, 49, 50, 51, 52, 55, 56, 58, 60, 62, 63, 65, 66, 68, 71, 74, 77, 81, 82, 85, 86, 89, 92, 94, 95, 97, 98, 105, 113, 114, 115, 116, 117, 118, 119, 120, 128, 134, 135, 139, 141, 143, 144, 145, 146, 147, 148, 149, 151, 152, 153, 155, 156, 157, 159
security challenges, v, viii, 19, 81, 82
security threats, 19, 21, 23, 52, 58, 114, 141, 156
settlements, 93, 98, 99, 100, 102, 103, 108, 109
social engineering, v, vii, ix, 113, 114, 115, 116, 117, 118, 119, 120, 121, 122, 123, 124, 125, 126, 127, 128, 129, 130, 131, 133, 134, 135, 137, 139, 141, 142, 143, 144, 145, 147, 149, 151, 152, 153, 155, 157, 159, 160
social environment, 85
social network, 141, 152
social welfare, 87
socio-cultural practices, 88
software, 2, 3, 4, 10, 16, 17, 19, 20, 21, 27, 28, 30, 32, 35, 37, 39, 57, 59, 61, 62, 63, 64, 67, 68, 116, 120, 121, 123, 139, 144, 145
stakeholders, 4, 87, 88, 100, 101, 102, 104, 105, 106, 107, 108
strategies, vii, viii, 66, 82, 83, 85, 86, 89, 98, 100, 105, 106, 158
structure, 7, 13, 14, 15, 16, 17, 18, 29, 34, 35, 38, 50, 51, 52, 53, 61, 62, 63, 100, 120, 121, 124, 125
success rate, 116

synchronization, 17, 18, 27, 53, 54
System Theoretic Process Analysis (STPA), vii, 1, 2, 3, 8, 9, 10, 32, 33, 36, 40, 49, 50, 51, 52, 55, 56, 58, 60, 63, 64, 65, 66, 67, 70, 72, 74, 76, 77, 78, 79
system theory, 2, 3, 9, 33, 34, 65, 152
systems dynamics, 114
Systems-Theoretic Accident Model and Processes (STAMP), vii, 1, 2, 3, 6, 8, 9, 10, 11, 32, 33, 34, 35, 36, 40, 49, 61, 62, 63, 64, 66, 67, 68, 70, 72, 74, 76, 77, 78

T

target population, viii, 82, 90
task performance, 150, 151
taxonomy, 116, 156
techniques, viii, 2, 6, 8, 9, 18, 32, 82, 144, 146
technological change, 88
technologies, vii, 1, 3, 4, 28, 29, 30, 33, 34, 61
technology, vii, viii, 1, 7, 17, 20, 28, 29, 32, 34, 35, 61, 68
threats, 2, 3, 7, 12, 15, 19, 20, 22, 25, 32, 52, 53, 55, 56, 58, 60, 66, 120, 139, 143, 145, 148, 153, 156, 157, 158

U

unacceptable risk, 11

V

validation, viii, ix, 1, 3, 29, 33, 39, 65, 114, 117, 136, 137, 155
villages, ix, 82, 84, 90, 97, 98, 106
virtual university, 111
vulnerability, 7, 8, 43, 44, 45, 50, 60, 97, 109, 110, 146
vulnerable and disadvantaged, 82

W

walking distance to school, 83
watering point, 93
weather patterns, 93, 97

Y

younger siblings, 92